pH and pION CONTROL
IN PROCESS
AND WASTE STREAMS

ENVIRONMENTAL SCIENCE AND TECHNOLOGY

A Wiley-Interscience Series of Texts and Monographs

Edited by ROBERT L. METCALF, *University of Illinois*

JAMES N. PITTS, Jr., *University of California*

WERNER STUMM, *Eidgenössische Technische Hochschulen*

PRINCIPLES AND PRACTICES OF INCINERATION
  Richard C. Corey

AN INTRODUCTION TO EXPERIMENTAL AEROBIOLOGY
  Robert L. Dimmick

AIR POLLUTION CONTROL, Part I
  Werner Strauss

AIR POLLUTION CONTROL, Part II
  Werner Strauss

APPLIED STREAM SANITATION
  Clarence J. Velz

PHYSICOCHEMICAL PROCESSES FOR WATER QUALITY CONTROL
  Walter J. Weber, Jr.

ENVIRONMENTAL ENGINEERING AND SANITATION
  Joseph A. Salvato, Jr.

NUTRIENTS IN NATURAL WATERS
  Herbert E. Allen and James R. Kramer, Editors

pH AND pION CONTROL IN PROCESS AND WASTE STREAMS
  F. G. Shinskey

pH AND pION CONTROL IN PROCESS AND WASTE STREAMS

F. G. SHINSKEY

Control Systems Consultant
The Foxboro Company

A WILEY-INTERSCIENCE PUBLICATION

John Wiley & Sons
NEW YORK • LONDON • SYDNEY • TORONTO

Copyright © 1973, by John Wiley & Sons, Inc.

All rights reserved. Published simultaneously in Canada.

No part of this book may be reproduced by any means, nor transmitted, nor translated into a machine language without the written permission of the publisher.

Library of Congress Cataloging in Publication Data:

Shinskey, F. Greg.
pH and pIon control in process and waste streams.

(Environmental science and technology)
"A Wiley-Interscience publication."
Includes bibliographies.
1. Factory and trade waste. 2. Process control.
3. Electrochemical analysis. 4. Hydrogen-ion concentration—Measurement. I. Title.
[DNLM: 1. Ions. 2. Water pollution—Prevention and control. WA689 S556p 1973]
TD897.5.S54   628'.3   73-7853
ISBN 0-471-78640-3

Printed in the United States of America

10 9 8 7 6 5 4 3 2 1

SERIES PREFACE
Environmental Science and Technology

The Environmental Science and Technology Series of Monographs, Textbooks, and Advances is devoted to the study of the quality of the environment and to the technology of its conservation. Environmental science therefore relates to the chemical, physical, and biological changes in the environment through contamination or modification, to the physical nature and biological behavior of air, water, soil, food, and waste as they are affected by man's agricultural, industrial, and social activities, and to the application of science and technology to the control and improvement of environmental quality.

The deterioration of environmental quality, which began when man first collected into villages and utilized fire, has existed as a serious problem since the industrial revolution. In the last half of the twentieth century, under the ever-increasing impacts of exponentially increasing population and of industrializing society, environmental contamination of air, water, soil, and food has become a threat to the continued existence of many plant and animal communities of the ecosystem and may ultimately threaten the very survival of the human race.

It seems clear that if we are to preserve for future generations some semblance of the biological order of the world of the past and hope to improve on the deteriorating standards of urban public health, environmental science and technology must quickly come to play a dominant role in designing our social and industrial structure for tomorrow. Scientifically rigorous criteria of environmental quality must be developed. Based in part on these criteria, realistic standards must be established and our technological progress must be tailored to meet them. It is obvious that civilization will continue to require increasing amounts of fuel, transportation, industrial chemicals, fertilizers, pesticides, and countless other products and that it will continue to produce waste products of all descriptions. What is urgently needed is a total systems approach to modern civilization, through which the pooled talents of scientists and engineers, in cooperation with social scientists and the medical profession, can be focused on the development of order and equilibrium to the presently disparate segments of the human environment. Most of the skills and tools that are needed are already in existence. Surely a technology that has created such manifold environmental problems is also capable of solving them. It is our hope that this Series in Environmental Science and Technology will not only serve to make this challenge more explicit to the established professional but that it also will help to stimulate the student toward the career opportunities in this vital area.

Robert L. Metcalf
James N. Pitts, Jr.
Werner Stumm

PREFACE

The decade of the seventies has been spoken of as the "age of pollution"—but I prefer to call it the "age of awareness," since the human family is finally becoming aware of the pollution that began with the smokestacks of the nineteenth century. A major reason for this new awareness has been the development of sensitive chemical analyzers. Sensitivity is the key to their success, because dangerous levels of poisonous substances polluting our air and water are measured in parts-per-million or parts-per-billion concentrations.

This book is devoted to the application of analytical devices for control of chemical ions in liquid streams. An example of the type of electrochemical analyzers used in these applications is the familiar pH measurement. To illustrate its sensitivity, a pH electrode can detect, with accuracy, hydrogen-ion concentrations at a level of 10^{-14} g-ions/l (pH 14). Converted to more familiar concentration units, 10^{-14} g-ions hydrogen/liter of water is 0.01 parts per quadrillion by weight. (The fact that the hydroxyl ion is in high concentration at pH 14 is beside the point—the electrode measures the concentration of hydrogen ions.)

By far the most common measurement made in stream-pollution control, particularly in the area of industrial waste treatment, is pH. Concentrations of acids and bases have a severe effect on aquatic life, and these materials are in very common use. As a consequence, the pH measurement has been important in waste treatment for many years. Fortunately, the very reliable and accurate glass electrode for measuring pH has been in use long enough that the procedures and practices of its application to pollution control have been thoroughly exploited.

More recently, an entire family of ion-sensitive electrodes has sprung forth—resembling, in principle, the pH electrode—but each sensitive to particular ion species. At first these ion-selective electrodes were greeted with feelings of: "Great, but where can they be used?" They have appeared on the marketplace at a very opportune time, however—just when public awareness to pollution by toxic ions is awakening. Now plants which have been discharging small quantities of mercury, lead, cyanide, sulfides, and other ions with their wastes, can monitor their concentrations. But more importantly, they now have the tools available to control the removal and recovery processes which convert these pollutants into valuable products or reusable feedstocks.

Other ion measurements such as reduction-oxidation potential are also made with electrodes directly inserted in the process stream. Many of the characteristics of the ion-selective electrodes apply to these measurements as well. In addition, they are also important in waste treatment: control of reduction-oxidation potential is an integral part of chromate and cyanide destruction.

These measuring devices have been developed largely in the laboratory by chemists, and later converted to process use by engineers. Their ultimate utilization, however, is in the on-line control of waste treatment and recovery process—and control is a long way removed from measurement.

To control a property of a process stream, an engineer must be knowledgeable in several areas. He must be familiar with the characteristics and limitations of the measuring devices, which in the case of ion electrodes, involves considerable physical chemistry. He must be familiar with the chemical reactions and the equipment in which they are carried out. Finally, he must be well grounded in control theory and application. A failing in any one of these areas will result in a system that is poorly controlled—and in this instance, poor control can bring not only destruction to life and equipment, but lawsuits and injunctions as well.

Since this wealth of knowledge and capability is not to be found readily in the average plant engineer or consultant, we bring it to you in one book. Part 1 is dedicated to the theory and practice of ion measurement, including equipment characteristics and failure modes. The processes in which these measurements are vital are outlined in Part 2. Most of the applications cited are well established waste-treatment processes. Some, however, are noted as relatively untried or even simply suggested uses for the newer ion-selective electrodes. In Part 3 you will learn how to combine the elements: electrodes, controllers, valves, reactants, vessels, pumps, and agitators, into a responsive coordinated system that accepts wastes and delivers useful products.

Control over ion concentration in waste streams is without doubt the most difficult problem in the field of process control. The facilities are expected to treat an unpredictable mixture of substances with widely differing properties. The variations in reagent load may range over several orders of magnitude. And the measuring devices are logarithmic and incredibly sensitive. If this were an easy problem, this book would not have to be written, for enough texts are already available on process control and its applications. But industrial waste control is exceedingly demanding, compared to other process-control problems, and the technology is needed *now*. So the author herewith presents his contribution to a more livable world.

The application of these methods and devices goes beyond stream-pollution control, however. Water treatment, beverage manufacture, ore flotation—in fact any process requiring an adjustment to the ion content of a stream—can be improved by more effective control. These processes are "child's play" compared to waste treatment—yet the principles and techniques for their control are identical. So, whereas the prime need for this book is in the field of waste treatment, the technology presented is equally applicable to any ionic process.

My sincere thanks go to Dr. Richard Oliver, my associate and consultant in matters of chemistry, and to my secretary Eileen Brennan who typed the entire manuscript with its many revisions. But my gratitude also extends to all those with whom I have worked, helping to solve their process-control problems. It is in these encounters with the problems of the world that we learn the most.

F. G. SHINSKEY

Foxboro, Massachusetts
March 1973

CONTENTS

PART 1 MEASUREMENT

CHAPTER 1   ELECTRODES AND THEIR CHARACTERISTICS   1

   Activity Measurements,   1

   The Nernst equation,   2
   The "p" notation,   4
   Inert electrodes,   5

   Reference Electrodes,   6

   The standard hydrogen electrode,   7
   The silver-silver chloride reference electrode,   7
   Solid-state electrodes,   12
   Calomel electrodes,   12
   The thalamid electrode,   13
   Liquid-junction potentials,   13

   Ion-Selective Electrodes,   15

   Glass-membrane electrodes,   15
   Solid-state electrodes,   18
   Liquid ion-exchange electrodes,   19

   Accuracy,   20

   Fundamental accuracy,   21
   Limits of detection,   22
   Interferences,   22

   References,   24

xi

CHAPTER 2 ION-MEASURING SYSTEMS   25

   The Transmitter, 25

   Potentials and resistances in the circuit, 26
   Ground paths, 29
   Temperature compensation, 33
   Adjustments, 36

   Electrode Assemblies, 37

   The submersible assembly, 38
   Flow-through assemblies, 41

   Maintenance, 43

   Transmitter calibration, 43
   Standard solutions, 43
   Standardization, 44
   Cleaning, 46

   References, 47

PART 2   APPLICATION

CHAPTER 3 THE HYDROGEN ION   51

   Strong Acids and Bases, 51

   Normality versus weight percent, 52
   Hydrogen ions from water, 53

   Weak Acids and Bases, 57

   Ionization of weak reagents, 57
   Neutralization of weak acids and bases, 60
   Buffering, 63
   Mixtures of strong and weak acids, 64
   "Strong" weak acids, 65

   Polyprotic Systems, 66

   Diprotic acids and bases, 66
   salts as acids and bases, 69
   Applications of hydrated lime, 74

Nonaqueous Media, 77

   Polar solvents, 77
   Nonpolar solvents, 78

Summary, 78

References, 79

CHAPTER 4 OTHER ION-SELECTIVE MEASUREMENTS   80

Monitoring, 80

   Effluent discharges, 81
   Leak detection, 81
   Column breakthrough, 82
   Total sulfide calculations, 83

Blending, 85

   Fluoridation, 86
   Water-hardness control, 87
   Scrubbing solutions, 88

Ion Removal, 92

   Solubility titration curves, 92
   Companion-ion control, 96
   Complexation, 99

References, 100

CHAPTER 5 REDUCTION-OXIDATION MEASUREMENTS   101

Reduction Reactions, 101

   Reduction potentials, 102
   Titration curves, 102

Oxidation of Cyanide Ions, 107

   Overall reactions, 107
   Hydrolysis of chlorine, 108
   Formation and hydrolysis of CNCl, 109
   Oxidation of cyanate ions, 110
   The plant and its controls, 112

Reduction of Chromate Ions, 113

> *Overall reactions,* 115
> *The reduction reaction,* 116
> *The oxidation reaction,* 117
> *The control system,* 119

Electrode Selection, 120

> *Noble metal electrodes,* 120
> *The antimony electrode,* 120
> *Special reference electrodes,* 122

References, 124

PART 3 CONTROL

CHAPTER 6 THE FUNDAMENTALS OF COMPOSITION CONTROL 127

Feedback Control, 127

> *Oscillation through phase shift,* 128
> *Loop gain,* 130

Dynamic Gain and Phase, 132

> *Dead time,* 132
> *Capacity,* 133
> *Combining dead time and capacity,* 135
> *Reaction rate lag,* 137

Steady-State Gain, 139

> *Electrode gain,* 140
> *Valve gain,* 142

Controllers, 145

> *Reset action,* 145
> *Derivative action,* 146
> *Combining the control modes,* 146

Summary, 152

References, 153

CHAPTER 7 DESIGNING A CONTROLLABLE PLANT 154

Mixing, 155

Blending, 155
Backmixing, 156
Choice of mixers, 160
Mixing and dynamic gain, 162

Vessel Selection, 164

Residence time, 164
Reaction rate, 166
Vessel arrangement, 169
Protection against failure, 172

Reagent Delivery, 174

Soluble reagents, 174
Insoluble reagents, 175

References, 180

CHAPTER 8 FEEDBACK CONTROL SYSTEMS   182

The Control Problem, 182

Accuracy, 183
Rangeability, 184
Nonlinearity, 186

Special Components, 187

Achieving wide rangeability, 188
The trim controller, 191
Valve characterizers, 191
The nonlinear controller, 194
Unsymmetrical functions, 198
Sampled-data control, 199

Performance, 200

Closed-loop response, 201
The effect of smoothing, 208
Multiple-stage neutralization, 209

Batch Processes, 210

Batch composition control, 211
Interrupted continuous operation, 213

References, 214

CHAPTER 9   FEEDFORWARD AND ADAPTIVE CONTROL   215

    Feedforward Control, 215

        Measuring flow, 216
        Determining normality, 218
        Narrow-range systems, 221
        Wide-range systems, 223

    Feedback Trim, 228

        The effects of weak agents, 228
        How to introduce feedback, 231

    Feedforward Adaptation, 232

        Variable valve gain, 233
        Variable titration curves, 235

    Feedback Adaptation, 236

        Another feedback loop, 237
        Assimilating performance, 238
        Adaptive-loop stability, 240

    Summary, 241

    References, 242

APPENDIX A. IONIZATION CONSTANTS OF ACIDS AND BASES   243

APPENDIX B. SOLUBILITY PRODUCT CONSTANTS   246

APPENDIX C. STANDARD REDUCTION POTENTIALS   248

APPENDIX D. TABLE OF SYMBOLS   251

INDEX   253

pH and pION CONTROL
IN PROCESS
AND WASTE STREAMS

PART 1

MEASUREMENT

1
ELECTRODES AND THEIR CHARACTERISTICS

Ions are charged particles drifting about in a polar solvent. Because they are charged, their very presence develops an electrical potential and their flow is also a flow of current. This book is dedicated to the control of ionic concentrations—but before they can be controlled, they must be measured.

Every polar solution contains more than one ion—for ionization of a chemical compound means separation into two oppositely charged particles. Water, for example, separates into hydrogen and hydroxyl ions. A chemical reactor may contain several different ions; a plant effluent may contain even more. The first step in achieving control of a reaction is finding the electrode combination which best describes its progress.

ACTIVITY MEASUREMENTS

A sensitive electrode develops a potential related to the *activity* of a certain ion or ions in the solution. Activity is related to but not the same as concentration:

$$a = \gamma x \qquad (1.1)$$

2 Electrodes

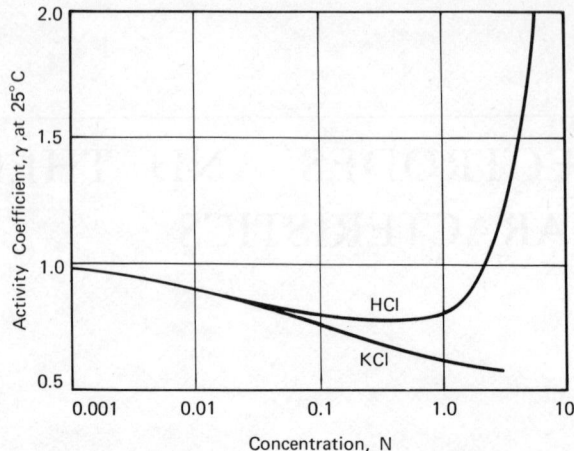

FIGURE 1.1. Hydrochloric acid is one substance whose activity coefficient passes through a minimum.[1]

where a is the activity expressed in the same units as concentration x of the ion, and γ is the activity coefficient. Units of concentration customarily used in ion analysis are gram-ions per liter of solution.

At infinite dilution, the activity coefficient approaches unity. With increasing concentration, γ begins to drop due to declining ion mobility. In many cases, further increasing the concentration of a solution raises γ above unity, due to a reduction in solvent concentration, particularly where hydration occurs. Figure 1.1 plots the activity coefficients of HCl and KCl versus concentration.

Most of the measurements of chemical systems are made in sufficiently dilute solutions that activity and concentration are essentially identical. But even where they are not, it is important to remember that activity is what brings about the chemical reaction as well as generating the electrode potential. So in most applications, *activity* is what we are really interested in measuring and controlling.

The Nernst Equation

The potential developed at a given electrode is called its half-cell potential, since two electrodes are needed for the complete circuit.

This potential is a function of the particular ionic species reacting with the electrode, and its activity. The potential is described in general terms by the Nernst equation:

$$E = E_0 + \frac{RT}{nF} \ln a \qquad (1.2)$$

when the electrode participates in the reaction. The coefficients of the logarithm are R, the universal gas constant [8314 mV-C/(°K)(g-mol)]; T, the absolute temperature in degrees Kelvin; n, the number of electrons deficient in the ion; and F, Faraday's constant (96,490 C/g-ion). E_0 is the unit potential of the half-cell, that is, the potential generated at unit activity, and is peculiar to each half-cell reaction. A table of these unit potentials appears in Appendix C.

Since it is more convenient to use base-10 than natural logarithms in expressing activities, the equation is usually given as

$$E = E_0 + 2.303 \frac{RT}{nF} \log a \qquad (1.3)$$

Substituting actual values for R and F, and selecting T at 25°C, (1.3) reduces to

$$E = E_0 + \frac{59.16}{n} \log a \qquad (1.4)$$

with E and E_0 expressed in millivolts. Observe that the potential change for univalent ions is 59.16 mV/decade, while for divalent ions it is 29.58 mV/decade. Positive ions increase potential while negative ions reduce it.

Example 1.1

The unit potential of a $Ag; Ag^+$ half-cell is $+797.8$ mV. The voltage it develops in a 0.1 N $AgNO_3$ solution at 25°C is

$$E = 797.8 + 59.16 \log 0.1$$

$$E = 797.8 - 59.16 = 738.6 \text{ mV}$$

The "p" Notation

One of the hurdles to overcome in potential measurements of ions is the logarithmic relationship between voltage and activity. This relationship pervades every phase of ionic measurement, from pH to reduction-oxidation potential. The most satisfactory method for working with these logarithms is to use the p (for power) notation as in pH:

$$\mathrm{pH} \equiv -\log a_{\mathrm{H}^+} \tag{1.5}$$

where a_{H^+} is the activity of hydrogen ions expressed in normality (g-ions/liter). Conversely,

$$a_{\mathrm{H}^+} = 10^{-\mathrm{pH}} \tag{1.6}$$

Unfortunately, the use of the p notation seems to have been limited to pH exclusively. However, it applies as well to other ions, for example,

$$\mathrm{pAg} \equiv -\log a_{\mathrm{Ag}^+}$$

The same principles can even be extended to equilibrium constants, which also range over many decades:

$$pK \equiv -\log K$$

Restating the Nernst equation in p units yields a linear form:

$$E = E_0 - \frac{59.16}{n}\, \mathrm{pIon} \tag{1.7}$$

Thus the pH scale is linear with voltage. The negative sign can be confusing, however, in that increasing activity is indicated by a decreasing pIon. The following illustrates one decade of activity:

pIon:	1.0	1.1	1.2	1.3	1.4	1.5	1.6	1.7	1.8	1.9	2.0
aIon:	0.100	0.080	0.063	0.050	0.040	0.032	0.025	0.020	0.016	0.013	0.010

Inert Electrodes

Platinum and gold electrodes are used to measure the condition of reduction-oxidation reactions taking place in a solution. Since the inert electrode does not take part in the reaction, the Nernst equation relates the ratio of the two species in equilibrium, that is, the oxidized and the reduced state of an ion. A substance is oxidized when it loses electrons, as:

$$\text{Fe}^{2+}(\text{reduced}) - e \xrightarrow{\text{oxidation}} \text{Fe}^{3+}(\text{oxidized})$$

The ferrous ion is "oxidized" to the ferric by losing an electron. Whether oxygen takes part in the reaction is inconsequential—chlorine, for example, is a strong oxidizing agent, and, in fact, supports combustion.

An inert electrode is sensitive to both these ions, developing a half-cell potential described by

$$E = E_0 + 2.303 \frac{RT}{nF} \log \frac{a_{\text{ox}}}{a_{\text{re}}} \qquad (1.8)$$

The potential E_0 is characteristic of a particular reacting pair such as Fe^{2+}, Fe^{3+}, and is developed when the activities of the two ions are equal. E_0 for Fe^{2+}, Fe^{3+} is $+760$ mV, and n is 1 since one electron is transferred per ion during the reaction. At 25°C, then, the potential of the ferrous-ferric half-cell is

$$E = 760 + 59.16 \log \frac{a_{\text{Fe}^{3+}}}{a_{\text{Fe}^{2+}}} \qquad (1.9)$$

Equation (1.9) can be rewritten using p notation:

$$E = 760 + 59.16(p\text{Fe}^{3+} - p\text{Fe}^{2+}) \qquad (1.10)$$

Thus far only one reactant and product have been touched upon. While one reactant is being oxidized, the other is being reduced—a similar half-cell potential then exists for the other agent and its corresponding product.

Chapter 5 is devoted entirely to common oxidation and reduction reactions whose state is measured with inert electrodes.

Observe that increasing the concentration of the oxidized state of the ion raises the potential and increasing that of the reduced state lowers it.

REFERENCE ELECTRODES

Reference electrodes are possibly the least understood and greatest source of trouble in ion-measuring systems. The first question to be asked is as follows: "Why are reference electrodes necessary?"

In order to measure the potential developed by ions in a solution, a closed circuit must be created, with *two* conductors in contact with the solution. A piece of metal inserted into a solution is sensitive indiscriminately to all the ions present, unless its own ions are there, or the metal is ionized by the solution. An inert electrode such as platinum, gold, or carbon cannot distinguish between chloride or hydroxyl ions, for example. But regardless of the choice of the first electrode, what will be used for the second? If the same material is used for both, then the same reactions will take place at both, yielding no potential difference between them.

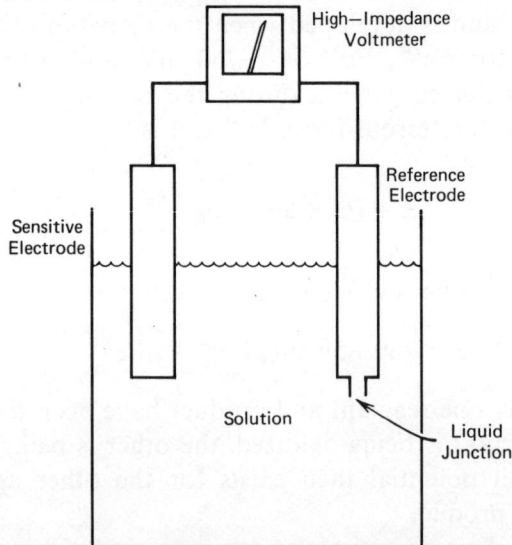

FIGURE 1.2. The reference electrode is designed to be insensitive to all ions in the solution.

This is where the reference electrode comes in—it is designed to be as *insensitive* as possible to all ions. Ideally its potential should be the same regardless of the solution in which it is placed. Much as a thermocouple needs a "cold junction" in an ice bath or some means of developing a similar reference potential, ionic measurements need a stable reference, too.

Figure 1.2 shows a pair of electrodes connected in a closed circuit. A high-impedance voltmeter is used to minimize the amount of current drawn through the circuit. More is said about this in the next chapter.

The Standard Hydrogen Electrode

The primary standard reference electrode is the standard hydrogen electrode (SHE). The electrode consists of a platinum wire or foil coated with platinum black, immersed in a solution of unit hydrogen-ion activity which is in equilibrium with hydrogen gas at atmospheric pressure. Its potential is set by convention to be 0 V at all temperatures. The SHE is not convenient to use for process applications or even for routine laboratory use, as it requires a supply of hydrogen gas, and is easily fouled in reducing solutions. Acceptable secondary reference electrodes are used in the process industry. The SHE is only mentioned at this time because electrode potentials are usually listed with reference to it.

Useful reference electrodes must produce a stable potential in any solution being measured, must be rugged, and insensitive to the environment. This actually limits the choice to very few. Two that are widely used in process work are the silver-silver chloride electrode and the mercury-mercurous chloride (calomel) electrode. Both are normally filled with a potassium chloride solution.

The Silver-Silver Chloride Reference Electrode

The silver-silver chloride electrode[2] is probably the most reproducible, reliable, and convenient reference electrode used in the process industries. It is easy to prepare, has less temperature hysteresis than the calomel electrode, and is nontoxic. In addition, its use with

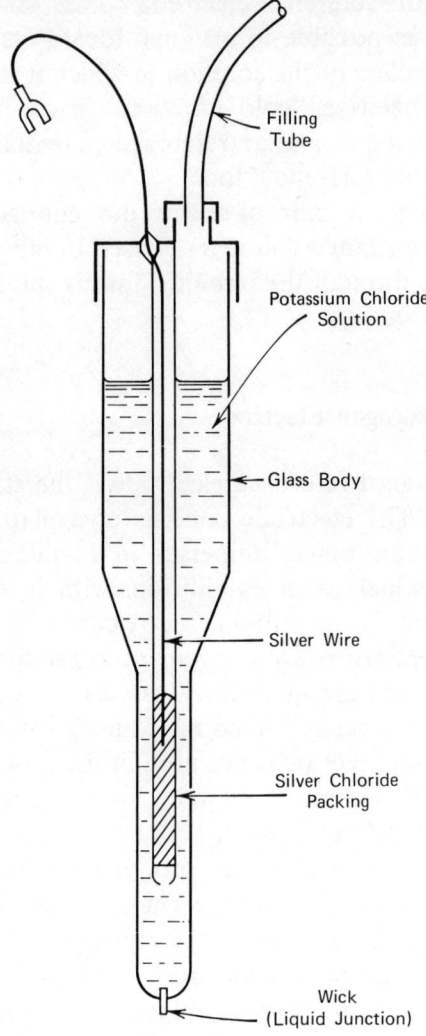

FIGURE 1.3. This reference electrode consists of an Ag, AgCl half-cell in a stable silver-ion environment.

sensitive electrodes having silver chloride internal cells results in a symmetrical circuit which minimizes temperature effects, and makes its compensation easier.

The silver-silver chloride electrode consists of a silver wire coated with insoluble silver chloride and placed in a solution of potassium chloride as shown in Fig. 1.3. The potassium chloride solution is saturated with silver chloride to prevent the dissolution of the silver chloride coating from the silver wire. The electrical contact between the process solution and the potassium chloride solution (the liquid junction) is made through a porous tip.

The silver-silver chloride electrode is itself an ion-selective measuring electrode. Its potential is a function of the silver-ion activity in solution as expressed by the Nernst equation:

$$E = E^\circ_{Ag, Ag^+} + \frac{2.3RT}{F} \log a_{Ag^+} \quad (1.11)$$

The activity of the silver ion is controlled by the solubility product (K_{sp}) of silver chloride (which is constant at a given temperature) and the activity of chloride ions in the solution:

$$K_{sp} = (a_{Ag^+})(a_{Cl^-}) \quad (1.12)$$

Solving for silver-ion activity,

$$a_{Ag^+} = \frac{K_{sp}}{a_{Cl^-}} \quad (1.13)$$

Substituting (1.13) into (1.11) gives

$$E = E^\circ_{Ag, Ag^+} + \frac{2.3RT}{F} \log \frac{K_{sp}}{a_{Cl^-}} \quad (1.14)$$

which rearranges to give

$$E = E^\circ_{Ag, Ag^+} + \frac{2.3RT}{F} \log K_{sp} - \frac{2.3RT}{F} \log a_{Cl^-} \quad (1.15)$$

10 Electrodes

The first two terms on the right-hand side of (1.15) can be combined to give

$$E = E^\circ_{Ag,AgCl} - \frac{2.3RT}{F} \log a_{Cl^-} \qquad (1.16)$$

where $E^\circ_{Ag,AgCl}$ is the unit potential of the silver-silver chloride half-cell. The potential of the silver-silver chloride electrode, therefore, is a function of the activity of the chloride ion in the reference solution. Potentials of several silver-silver chloride electrodes in different strengths of potassium chloride solution at 25°C are shown in Table 1.1, with reference to the SHE at zero.

The saturated potassium chloride electrode is most commonly used because its solution is easy to prepare. However, the solubility of potassium chloride changes with temperature, and this can create difficulties when the reference electrode is used in a process subjected to variations in temperature. Table 1.2 shows the solubility of potassium chloride in water at various temperatures from 0 to 100°C. It may be noted that the 4 M KCl electrolyte is saturated at 19.4°C while a saturated potassium chloride solution at 25°C is 4.17 M. The solubility of potassium chloride increases nearly twofold over the temperature range of 0 to 100°C. If the common 4 M potassium chloride reference electrode is used where ambient temperatures fall below 19.4°C, crystals will form, falling into the tip of the electrode, and possibly plugging the liquid junction.

In water, silver chloride is considered an insoluble salt. However, in concentrated potassium chloride solutions, a soluble complex ion

Table 1.1 Potentials of Silver-Silver Chloride Electrodes at 25°C

KCl Concentration (M)	$E_{Ag,AgCl}$ (mV)
0.1	+293.8
1.0	235.8
4.0	199.8
4.17 (saturated)	198.5

Table 1.2 Solubility of Potassium Chloride

Temperature (C)	Concentration (M)
0	3.39
19.4	4.00
20	4.02
25	4.17
50	4.80
100	5.84

is formed. The data in Table 1.3 show that the solubility of silver chloride in 4 M potassium chloride solution at 25°C is 913 mg/l—this is 65 times more soluble than in a 1 M potassium chloride solution! Whenever the electrode cools, silver chloride will crystallize in the body and in the tip. The silver ion from the dissolved silver chloride may also form an insoluble precipitate at the electrode tip if the process solution has ions such as sulfide, carbonate, or phosphate that form less soluble silver salts. Because of its greater solubility, much more silver chloride is available for precipitation from a 4 M potassium chloride electrolyte than from a 1 M potassium chloride solution. In a saturated potassium chloride electrolyte, the solubility of silver chloride is even greater.

Since few processes operate at constant temperature, the crystallization problem must be faced. This fact, plus the increased stability demands of ion-selective electrode applications, makes it worth-

Table 1.3 Solubility of Silver Chloride in Potassium Chloride Solutions at 25°C

KCl Concentration (M)	AgCl Solubility (mg/l)
0.100	0.40×10^{-3}
1.00	14.4
4.00	913
4.17	1000

while to use a lower concentration of potassium chloride, such as 1.0 M. The upper temperature limit of this electrode is 110°C, at which point the silver chloride starts to decompose. The lower temperature limit is the freezing point of the solution, -11°C. One-molar KCl may freeze homogeneously in its reservoir while it still may be liquid at the junction and perform satisfactorily. The 4 M solution precipitates both KCl and AgCl before freezing, affecting calibration and tending to plug the liquid junction.

Solid-State Electrodes

Recently certain "solid-state" reference electrodes have come into use, which do not have a flowing internal solution. They are silver-silver chloride electrodes containing a gel saturated with KCl and AgCl. The liquid junction to the process is made through a porous medium such as wood, ceramic, or polymer. Because the cavity of the electrode is completely filled with gel, external pressure cannot easily force process fluid into the electrode to contaminate it. Stability is given as ± 6 mV (± 0.1 pIon) per year, acceptable for most applications. Low maintenance is their principal contribution. Their half-cell potential is that of the 4.17 M electrode given in Table 1.1.

Calomel Electrodes

Another reference electrode commonly used in the process industries is the saturated calomel electrode (SCE). This electrode consists of a layer of mercury covered with a paste of mercury, mercurous chloride (calomel), and potassium chloride, all in contact with a solution saturated both in potassium chloride and mercurous chloride.

It can be demonstrated by an argument similar to that for the silver-silver chloride electrode that the potential of a calomel electrode is dependent on the concentration of chloride ion in the reference solution. The SCE has a potential of 241 mV when compared with the SHE at 25°C. The normal calomel electrode (NCE), a common laboratory reference electrode, contains a 1.0 N

solution of potassium chloride, and exhibits a potential of 280 mV at 25°C. Calomel electrodes are more in favor in Europe, while silver chloride electrodes are used almost exclusively in the western hemisphere.

The Thalamid Electrode

The thalamid electrode is similar to the calomel electrode, consisting of a thallium amalgam, thallous chloride, and saturated potassium chloride. The advantage of the thalamid electrode is that it can be used up to 140°C. Its major disadvantage is its potential (-572 mV vs. SHE at 25°C) which results in a serious mismatching between reference and measuring electrodes.

Liquid-Junction Potentials

When two solutions of different composition are brought into contact, their ions tend to diffuse from the more concentrated into the more dilute solution. However, different ions tend to diffuse at different rates—this characteristic is known as "ionic mobility." The hydrogen ion travels most rapidly, having an ionic mobility of 0.00362 cm/(sec)(V)(cm) at 25°C. The mobility of the hydroxyl ion is 0.00205, while that of the potassium ion is 0.00076, and of the chloride ion is 0.00079.[3]

The difference in the diffusion rates between a cation and an anion develops a charge known as the *liquid-junction potential*. In every ion-measuring system containing a liquid junction, this potential must be taken into account.

Potassium chloride has been chosen as a filling solution for reference electrodes because the mobilities of K^+ and Cl^- ions are nearly identical. When the reference solution comes in contact with a solution of HCl, however, the higher mobility of the H^+ ion causes a negative junction potential proportional to its activity. By the same token, contact of a KCl solution with NaOH will produce a positive junction potential due to the higher mobility of the OH^- ion.

14 Electrodes

Table 1.4 Liquid-Junction Potentials between 4 M KCl and Various Solutions at 25°C

Solution	E_j (mV)
HCl, 1 M (pH 0)	−14.0
HCl, 0.1 M (pH 1)	−4.5
HCl, 0.01 M (pH 2)	−2.9
KH Phthalate, 0.05 M (NBS Standard pH 4.01[a])	−2.5
KCl, saturated (pH 7)	+0.1
KCl, 1.0 M (pH 7)	−0.7
KH$_2$PO$_4$, 0.025 M; Na$_2$HPO$_4$ 0.025 M (NBS Standard pH 6.87[a])	−1.8
NaHCO$_3$, 0.025 M; Na$_2$CO$_3$, 0.025 M (NBS Standard pH 10[a])	−1.7
NaOH 0.01 M (pH 12)	−2.2
NaOH 0.10 M (pH 13)	+0.5
NaOH 1 M (pH 14)	+8.7

[a]NBS includes liquid-junction potentials when assigning pH values to standards.

Table 1.4 lists the junction potentials developed by a 4 M KCl solution in contact with various process solutions.

Liquid-junction potentials can be eliminated by using a reference electrode without a liquid junction or by using a salt bridge with filling solution the same as the process solution. Ion-selective measuring electrodes, including the pH electrode, have been used as a reference electrode on certain specific applications. The only requirement is that the solution being measured has a constant activity of the appropriate ion or that a constant activity of the ion can be added. Unfortunately, neither of these conditions is commonly met in process solutions.

When a reference electrode is combined in a circuit with a sensitive measuring electrode as in Fig. 1.2, a voltage difference is developed:

$$E = E_{\text{meas}} - E_{\text{ref}} + E_j \qquad (1.17)$$

If the junction potential is significant, the electrodes may be standardized in a buffer more nearly representative of the process solution.

ION-SELECTIVE ELECTRODES

Up to this point, we have examined inert electrodes which respond to all ions and reference electrodes which respond to none. But recall that the half-cell of the silver-silver chloride electrode was sensitive to the activity of silver ions, and indirectly to the activity of chloride ions, in the filling solution. The silver wire coated with silver chloride has been used as a silver-ion electrode in laboratories for years. It is fragile, however, in that the coating may be removed by abrasion or replaced by some less soluble salt. Recently, an entire family of ion-selective electrodes[4] (including a silver-ion electrode) has been developed for industrial use. Their general properties are given below, while their applications are described in Chapters 3 and 4.

Glass-Membrane Electrodes

Glass electrodes are constructed from specially formulated glass membranes that respond to ions by means of an exchange of mobile ions within the membrane structure. Figure 1.4 is a diagram of the conventional pH electrode, identifying three components essential to membrane electrodes: the membrane, the internal filling solution, and the internal reference electrode. The membrane is fused to a glass body so that the outer surface makes contact with the process while the inner surface makes contact with the internal filling solution. The latter contains a constant activity of hydrogen ion while the process solution contains an unknown activity of the same ion. Changes in the membrane potential are proportional to changes in the process pH. The change is measured by providing a stable electrical connection with the internal solution.

Some commercial electrodes employ metal-metal ion systems such as mercury immersed in mercurous perchlorate and perchloric acid. But the greatest number of glass electrodes are filled with solutions of hydrochloric acid or buffered chloride into which a silver-silver chloride electrode is immersed. This electrode is known for its high thermal stability, day to day reproducibility, and low hysteresis after changes in temperature.

16 Electrodes

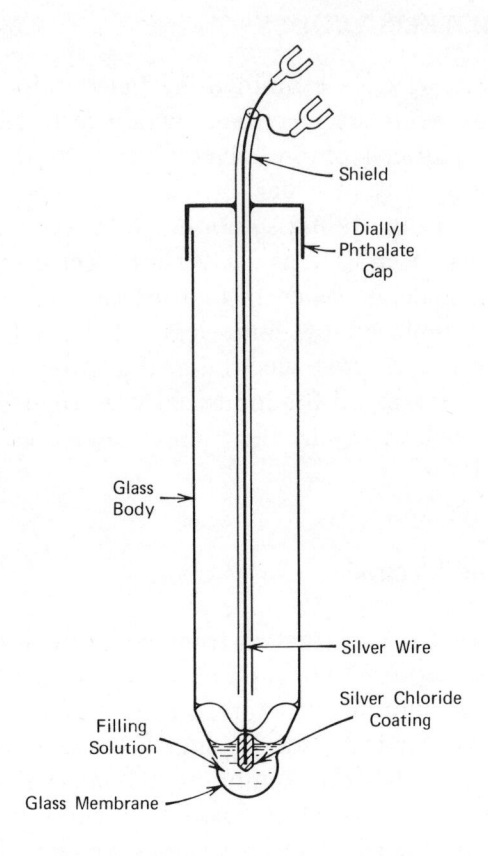

FIGURE 1.4. The glass pH electrode contains a silver-silver chloride half-cell and a filling solution buffered to pH 7.

Metal-filled glass electrodes and metal films deposited on the inner surface of the membrane have also been used to provide the inner electrode contact. However, this technique has not found wide use. While the silver-silver chloride internal is a good reference element, it has the same temperature limitations as the external reference electrode.

Whatever the internal reference element, other than a metal film, there must be an internal solution containing a constant activity of the ion for which the internal element is responsive. This ensures a stable potential at the internal contact. Also the solution must contain a constant activity of hydrogen ion to establish the inner potential contribution for the membrane. In the conventional pH electrode, the solution is buffered at pH 7 and contains a chloride level similar to that used in the external reference electrode (4 M KCl).

When the internal half-cells of the measuring and reference electrodes are identical, their voltage contributions cancel one another. The entire potential difference in the circuit is then developed across the membrane itself, between the activities of the ions on either side:

$$E = \frac{2.3RT}{nF} \log \frac{a_{\text{process}}}{a_{\text{fill}}} \qquad (1.18)$$

For the pH electrode with a filling solution buffered at 7, the potential is

$$E = \frac{2.3RT}{F} \log \frac{a_\text{H}}{10^{-7}}$$

or, expressed in p units,

$$E = \frac{2.3RT}{F}(7 - \text{pH}) \qquad (1.19)$$

At pH 7, there is no temperature error—this is known as the isoelectric point. The only errors likely at pH 7 are asymmetry errors—slight differences between the two surfaces of the membrane.

Other glass electrodes finding process use are the Na^+, NH_4^+ and K^+ electrodes. These have a construction similar to that of the pH electrode, differing primarily in the composition of the glass membrane and internal filling solution. Glass electrodes can also be constructed by slicing thin sections from a rod of glass and cementing them to an epoxy body. This eliminates the familiar glass body. The epoxy construction is available for the sodium electrode but not yet available for pH electrodes due to difficulties inherent in cementing pH glass to epoxy.

Dissolved-gas sensing electrodes for CO_2 and NH_3 can be made from conventional pH electrodes. The glass membrane is covered by a permeable membrane sac filled with a carbonate or ammonium

18   Electrodes

buffer. The appropriate gas in equilibrium with the solution will selectively diffuse through the permeable membrane, creating a change in the pH of the buffer solution. This change in pH is related to the activity of the gas in equilibrium with the process solution. These electrodes do not appear to be in use in process control but there is no reason why they could not be. An especially interesting application would be the control of CO_2 in carbonated beverages.

Solid-State Electrodes

Solid-state electrodes are made from single or polycrystals of insoluble conductive compounds. The crystalline character of the compound serves to restrict the size and charge of the ions which can move within the lattice network.

The composition of the lattice varies as a function of the required measurement. The fluoride electrode, for instance, has a single crystal of lanthanum fluoride for a sensing membrane. The silver and sulfide membranes are pressed pellets of insoluble silver sulfide. The solubility of silver sulfide is so low that it prevents the coexistence of silver and sulfide ions (except in extremely small amounts)

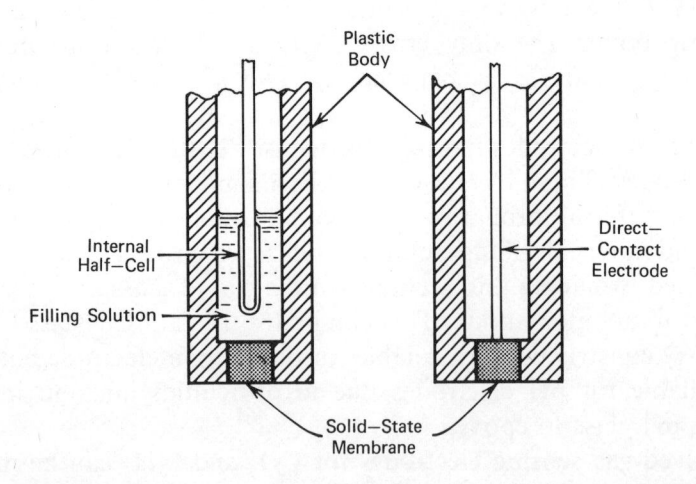

FIGURE 1.5. Solid-state electrodes may contain an internal half-cell or direct internal contact.

Table 1.5 Solid-State Electrodes and Membrane Composition

Electrode	Membrane	Form
Fluoride	LaF_3	Single crystal
Silver/sulfide	Ag_2S	Pressed pellet
Chloride, bromide, or iodide	$AgX-AgS$	Pressed pellet
Cyanide	$AgI-Ag_2S$	Pressed pellet
Copper	$CuS-Ag_2S$	Pressed pellet

and the electrode can be used to measure either of these ions. Like the sodium electrode mentioned above, these membranes are sealed in epoxy bodies, as shown in Fig. 1.5.

Table 1.5 lists some of the commercially available solid-state electrodes and the composition of the sensing membrane. Some pressed pellets and the single crystalline silver-salt membranes are capable of having a metal deposited on the surface and an electrical lead connected to the metal deposit. A solid connector permits the use of the electrodes in any position without breaking electrical continuity. Also, there are no internal solutions which could deteriorate with time or temperature.

Liquid Ion-Exchange Electrodes

There are many ions for which no glass or crystalline membrane can be found that is suitable for process measurements. Fortunately, by using techniques familiar in ion-exchange and solvent-extraction technology, electrodes can be built for some of these ions. An inert, hydrophobic membrane, such as a treated filter paper, ceramic, or organic film can be made selective to certain ions by saturating it with an organic ion-exchange material dissolved in an organic solvent. Figure 1.6 shows how such an electrode is assembled. This electrode has two filling solutions, an internal aqueous filling solution in which the silver-silver chloride reference electrode is immersed, and an ion-exchange reservoir of a nonaqueous water-immiscible solution, which saturates the porous membrane. The membrane serves only as a support for the ion-exchange liquid,

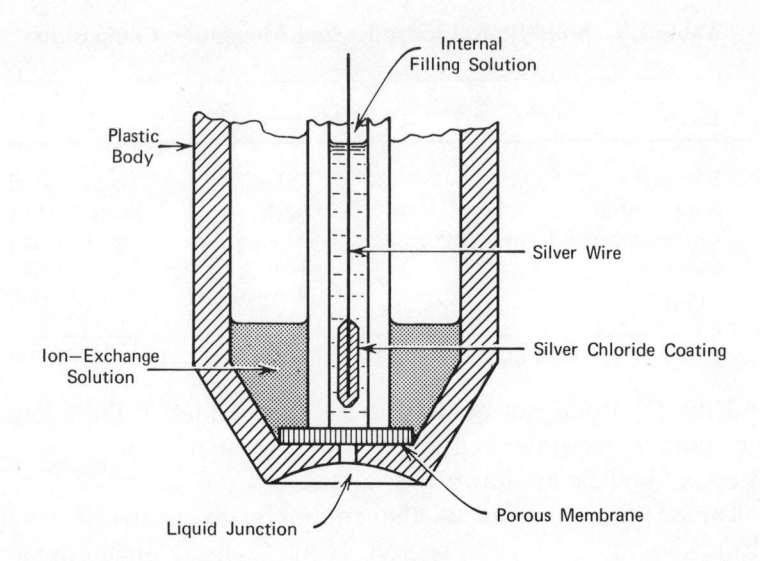

FIGURE 1.6. The liquid ion exchange electrode has two filling solutions; the ion exchanger must be replenished.

separating the internal filling solution from the unknown solution in which the electrode is immersed. If the liquid ion exchanger is selective for calcium, for example, a potential across the membrane is created by the difference in calcium activity between the internal filling solution and the process solution. In this case, the internal filling solution contains a constant activity of calcium ions.

The electrode is designed so that the liquid ion exchanger, used as a sensing element, has a very small positive flow into the process stream. Liquid ion-exchange membrane electrodes, therefore, require recharging with ion exchanger. A single recharging should last several months in a properly designed system. Unlike the solid-state or glass electrodes, liquid-membrane electrodes cannot be used in nonaqueous solutions which would dissolve the liquid ion exchanger.

ACCURACY

It is not often easy to resolve a disagreement between a process ion measurement and a laboratory analysis of the same material. Gravimetric and spectrophotometric analyses and titrations generally report results in concentration units such as milligrams per liter. In fact, this is the form demanded by pollution-control authorities. By the same token, activity measurements such as pH are equally important, although they will not often agree with a thorough laboratory analysis. Activity and concentration are not the same, as (1.1) indicates. But beyond that, much of an element in a given sample may not be ionized—if all the hydrogen in water were ionized, it would be 55 N acid! Molecules may fail to ionize either because they are tightly bound as in a weak acid, because they have been precipitated, or converted to a gas, or complexed by some other agent. All of these situations are discussed in detail in later chapters. At the moment, we are interested in how accurately a given electrode is able to indicate the activity of the ions actually surrounding it.

Fundamental Accuracy

Even with very carefully calibrated instruments and standardized electrodes, measurements of ion activity probably cannot be made more accurately than ±1 mV. And to achieve this degree of precision requires standardization against a known sample similar to the process solution, or otherwise taking into account liquid-junction potentials (which Table 1.4 indicates may be several millivolts).

At roughly 60 mV/decade, ±1 mV amounts to about ±1.7% of a decade or ±0.017 pIon units. To convert this error into activity or concentration units, the logarithmic relationship must be differentiated:

$$\text{pIon} = -\log a$$

$$\frac{d\text{pIon}}{da} = -\frac{\log e}{a} = -\frac{0.434}{a}$$

Then, the activity error da can be related to the pIon error:

$$\frac{da}{a} = -2.3\, d\text{pIon}$$

$$\frac{da}{a} = -2.3(\pm 0.017) = \pm 0.038$$

A 1-mV error corresponds to an activity or concentration error of ±3.8% of value. At a pIon of 1.0, for example, activity would be 0.1 $N \pm 0.0038$ N; at a pIon of 2.0, activity would be 0.01 $N \pm 0.00038$ N.

An accuracy of ±3.8% of value is not particularly good for an analytical instrument, and this is mainly due to the logarithmic nature of the measurement. However, this same characteristic provides tremendous rangeability and sensitivity, which more than make up for the limitation of absolute accuracy.

Limits of Detection

The upper limit of detection for ion-selective electrodes is the saturated solution. However, due to the problems of making measurements with reference electrodes having large liquid-junction potentials (see Table 1.4), electrodes are specified as having an upper limit of 1 M. If the problems of large liquid junction are brought under control, measurements can be made in saturated or nearly saturated solutions. The lower limit of detection is usually determined by the solubility of the solid-state sensing element or the liquid ion exchanger. The solution pH sometimes determines the lower limit of detection. Some dilute solutions are unstable, but activity measurements may be made if the solution is buffered with respect to the ion being measured; that is, if the free ion is in equilibrium with a relatively large excess of complexed ion. This is the case when measuring free silver in photographic emulsions, or sulfide, cyanide, or fluoride in acid solutions.

Interferences

All ion-selective electrodes are similar in their principle of operation and in the way they are used. They differ only in the details of the process by which the ion to be measured moves across the membrane and by which other ions are kept away. Therefore, electrode interferences are a function of the membrane materials.

The glass electrodes and the liquid ion-exchange electrodes both function by means of an ion exchange of mobile ions within the membrane, and ion-exchange processes are not specific. Reactions will occur among many ions having similar chemical properties such as the alkali metals, alkaline earths, or transition elements. Thus a number of ions may produce a potential when a given ion-selective electrode is immersed in a solution. Even the pH glass electrode will respond to sodium ions at a very high pH (low hydrogen-ion activity). Fortunately, an empirical relationship is available which can be used to predict electrode interferences, and a list of selectivity ratios for the interfering ions can be obtained by consulting the manufacturers' specifications or the chemical literature.

Solid-state electrodes are made of crystalline materials, and interferences resulting from ions moving into the solid membrane are not to be expected. Interference is usually by a chemical reaction with the membrane. One which is observed with the silver halide membranes (for chloride, bromide, iodide, and cyanide activity measurements) involves reaction with an ion in the process solution (such as sulfide) which forms a more insoluble silver salt. As above, specific details of electrode side reactions can be found in the manufacturers' specifications and chemical literature.

A true interference is one that produces an electrode response which can be interpreted as a measure of the ion of interest. For example, the hydroxyl ion (OH^-) causes a response with the fluoride electrode at levels of fluoride below 10 ppm. Also, the hydrogen ion (H^+) creates a positive interference with the sodium-ion electrode. Often an ion will be thought of as an interference if it reduces the activity of the ion of interest through chemical reaction. It is true that this chemical reaction (complexation, precipitation, oxidation-reduction, hydrolysis, etc.) results in an activity of the ion which differs from the concentration of the ion by an amount greater than that caused by ionic interactions. However, in this case the electrode is still measuring the true activity of the ion in the solution.

Sometimes this interference can be beneficial. For example, the Ca^{2+} electrode can be made sensitive to other divalent ions such as Ba^{2+}, Mg^{2+}, Zn^{2+}, Ni^{2+}, and Fe^{2+}—all of which make water "hard." This electrode is called the divalent cation electrode, to distinguish it from the calcium-ion electrode, and is used for water-hardness measurements.[5] Its internal filling solution contains calcium ions, so that hardness is essentially reported in terms of Ca^{2+} activity.

REFERENCES

1. F. Daniels, *Outlines of Physical Chemistry*, John Wiley & Sons, Inc., New York, 1948, p. 473.
2. "Ion-Selective Reference Electrodes," *Technical Information Sheet* 43-33a, The Foxboro Company, Foxboro, Massachusetts, March 1971.
3. F. Daniels, *op. cit.*, p. 414.
4. "Ion-Selective Measuring Electrodes," *Technical Information Sheet* 43-32a, The Foxboro Company, Foxboro, Massachusetts, March 1971.
5. "Water-Hardness Measuring System," *Technical Information Sheet* 43-34a, The Foxboro Company, Foxboro, Massachusetts, September 1971.

2

ION-MEASURING SYSTEMS

Having developed a potential difference as great as several hundred millivolts between a selected pair of electrodes, the next task is to find an instrument capable of measuring that potential. This instrument must be sensitive, accurate, flexible enough to use with many combinations of electrodes, and fully adjustable. It must be capable of transmitting a signal proportional to the potential difference over several hundred feet to recording and control devices with a minimum sensitivity to induced error and electrical interference.

The electrodes themselves must be mounted in a weatherproof assembly, protected from the process, yet easily removed for maintenance. Cables and connections must be of high quality to protect against open and short circuits. All of these components fit together to form a measuring "system." Each has its own contribution to system performance, and each is discussed individually in the pages that follow.

THE TRANSMITTER

The device which converts the millivolt potential difference from the electrodes into a signal with a standard range of 4 to 20 mA, may be called a transmitter, a pH-to-current converter, or simply an amplifier. Its most distinguishing feature is an extremely high input

26 Measuring Systems

impedance, since the membrane of the glass electrode may have a resistance exceeding 500 MΩ. An equivalent circuit will help point out the requirements for such an instrument.

Potentials and Resistances in the Circuit

The pH measuring system will be used as an example throughout the discussion on circuits, since it is the most common and also because it features extremely high membrane resistance and a midscale isopotential point. Figure 2.1 shows the equivalent circuit of pH measuring and reference electrodes, a solution, and a meter. (Conducting paths to ground have been omitted for simplicity, to be examined later in the chapter.) The sources of voltage in the loop include the matched Ag, AgCl half-cells in both electrodes, the variable potential E across the glass membrane, and the liquid-junction potential E_j. If the junction potential is insignificant, the

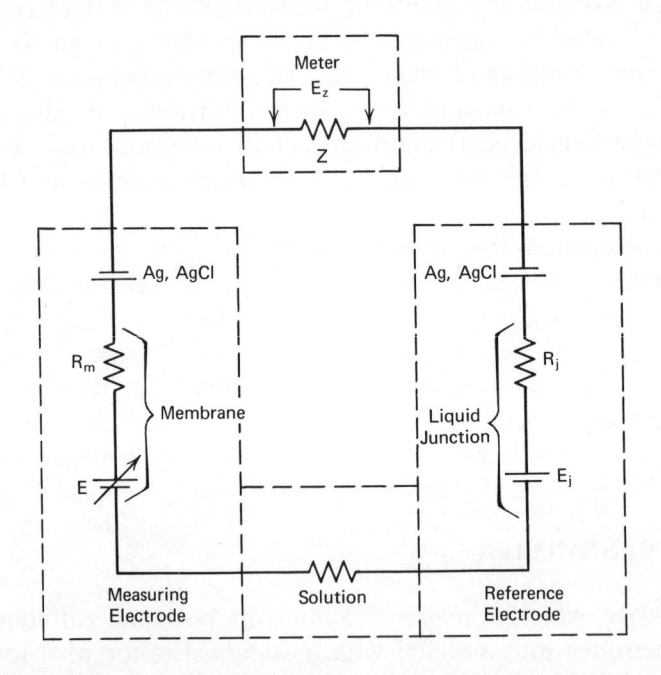

FIGURE 2.1. This equivalent circuit shows only those elements in the normal path of current flow.

net electromotive force to be measured is E, related to the solution H^+ activity:

$$E = \frac{2.3RT}{F} \log \frac{[H^+]_{process}}{[H^+]_{fill}} \quad (2.1)$$

where the brackets indicate activity of the species they enclose.

This potential develops the indicated voltage E_Z across the meter. However, the resistances in the loop have a pronounced effect on the relationship between E_Z and E. They consist of the resistance of the reference solution across the liquid junction, that of the solution between the electrodes, the resistance of the membrane, and finally, the impedance Z of the meter. The resistance of the liquid junction is ordinarily about 10,000 Ω, and compared with that of the membrane, is negligible, unless the junction is plugged or otherwise contaminated.

Solution resistance is ordinarily in the same range or even less, and is for most applications, inconsequential. There are situations, however, where it may be extremely significant. Deionized or distilled water has a resistance of hundreds of megohms which can cause considerable difficulty in making a reliable ion measurement. Nonaqueous media such as benzene or methanol can cause similar impedances unless ionic impurities are present. Solution resistance can also be increased by moving the electrodes apart. To measure pH successfully in a poorly conducting solution, the electrodes should be mounted as close together as possible, with the reference electrode upstream of the glass electrode. A high-flow liquid junction will then allow a stream of KCl to form a conducting path between the electrodes.

All the resistances in the circuit are in series, so that membrane, junction, and solution resistances may be added and compared with the meter impedance. Figure 2.2 is a reduction of Fig. 2.1 to those elements. Notice that the circuit is reduced to a simple voltage divider, wherein the meter voltage is a portion of the cell voltage:

$$E_Z = E \frac{Z}{Z+R} \quad (2.2)$$

The error caused by Z being less than infinite or R being other than

28 Measuring Systems

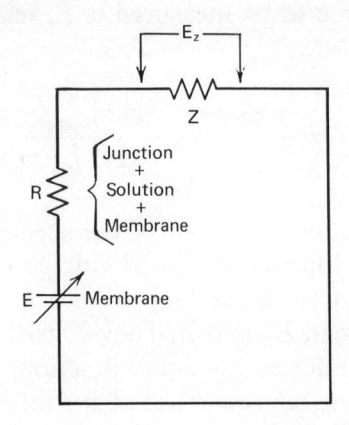

FIGURE 2.2. A simplified representation of the normal signal path.

zero is the difference ΔE between the actual and measured potential:

$$\Delta E = E - E_Z = E - E\frac{Z}{Z+R}$$

The error is relative to the actual cell potential:

$$\Delta E = E\frac{R}{Z+R} \qquad (2.3)$$

Bear in mind that the pH electrode is filled with a buffer such that E is zero at pH 7. Impedance errors are therefore minimized in this region.

Since pH is linear with E, (2.3) could be expressed in terms of pH error (ΔpH):

$$\frac{\Delta \text{pH}}{\text{pH} - 7} = \frac{R}{Z+R} \qquad (2.4)$$

A 1% error in pH relative to the departure from 7 would require Z to be 100 times as great as R.

Membrane resistance is far from constant, so that the impedance-induced error varies, too. The most pronounced influence is that of temperature, as shown in Fig. 2.3. With measuring instruments having input impedances typically around 100,000 MΩ, errors due to membrane resistance are not likely to be significant above 20°C.

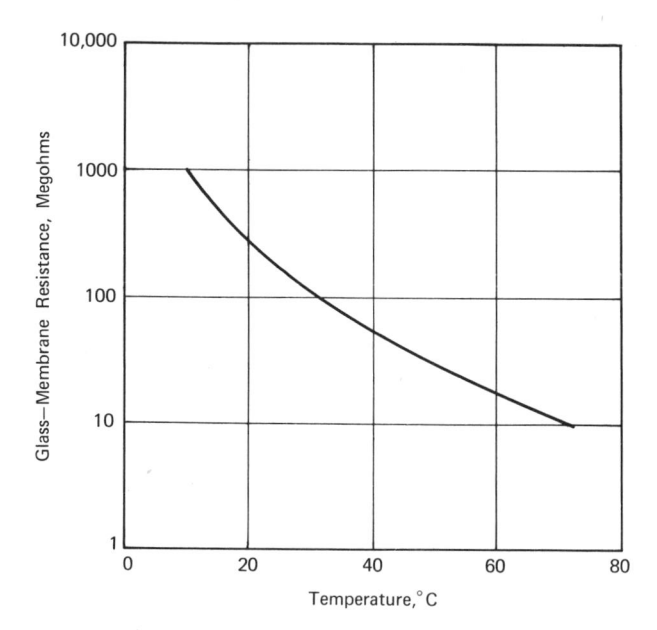

FIGURE 2.3. The membrane resistance of Corning 015 glass.[1]

Membrane resistance is also increased by drying. A dry electrode may require as much as an hour's immersion in water before its resistance stabilizes. Similarly, anhydrous nonaqueous media may raise its resistance to unusable levels by dehydration, in absence of trace amounts of moisture.

Ground Paths

If measurements did not have to be made on solutions in process vessels, many of the failures encountered in these circuits could be avoided. But the solution also conducts current to the vessel and its piping, which may be grounded in any number of places.[2] Figure 2.4 shows possible paths to ground at the electrode terminals as well as in the solution. Obviously any paths to ground within the measuring instrument or connecting cables can be equated with either of those designated as terminal grounds in the figure.

Observe first that terminal resistances R_1 and R_2 effectively bypass the instrument impedance. They could be low due to a film

30   Measuring Systems

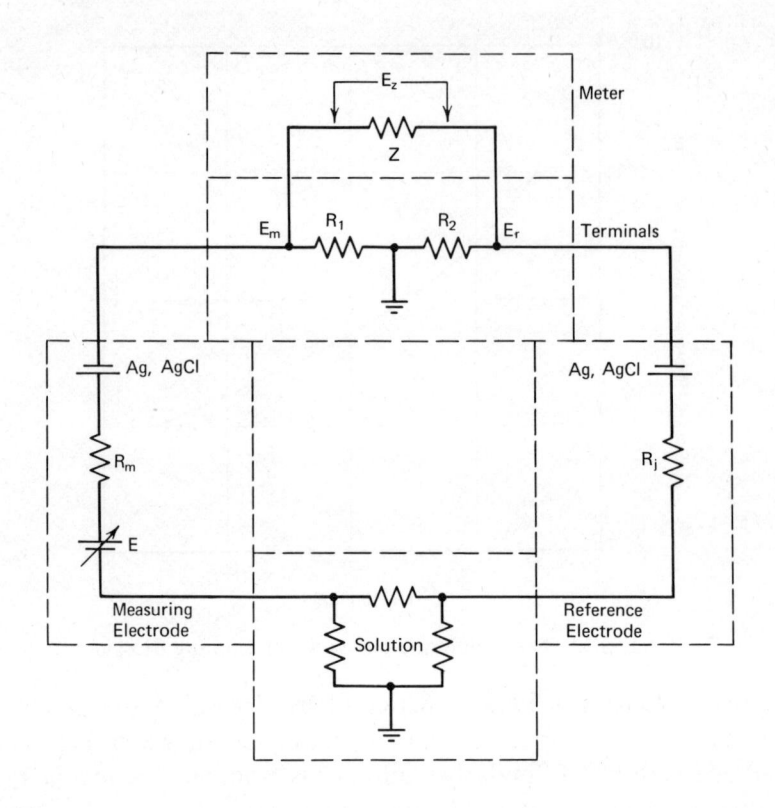

FIGURE 2.4. Current may flow between the terminal connections and the solution through ground paths.

of moisture across the insulating barrier separating the terminals, or simply due to high humidity or water accumulated anywhere in the wiring. A piece of ordinary electrical tape placed between the electrode leads is a resistance low enough to cause considerable error. Manufacturers employ insulation materials of low moisture adsorption, such as diallyl phthalate, to keep these resistances high. Since the terminal strip is mounted in some sort of junction box, the same considerations must be given terminal resistance to the box itself, that is, ground.

A low resistance to ground from the reference electrode is less serious than that from the measuring electrode. To illustrate, assume the solution is grounded directly, and that R_1 and R_2 are both 10 MΩ. Reference potential E_r will be reduced by the ratio of the reference-electrode resistance R_j to the total in that circuit, $R_j + R_2$. If R_j is only 10,000 Ω, E_r will be reduced by only 0.1%. Applying the same terminal resistance to the measuring electrode with a membrane resistance R_m of 10 MΩ, would result in a 50% reduction in measurement potential E_m.

The effect of R_1 does not depend on R_2 since a closed circuit is formed through R_1 and ground to the solution. Consequently great care must be exercized in wiring the measuring electrode, whereas the reference electrode is several orders of magnitude less sensitive.

A most important conclusion to be drawn from this discussion is the probable failure mode of a measuring system. A low resistance from the measuring electrode to ground or to the reference electrode will cause a reduction in meter voltage, indicating a pH more nearly 7 than it should be. A pH reading of 7 does not necessarily indicate proper operation—it could also indicate a dead short!

Every pH measuring instrument has a function switch with a "standby" position. When the switch is in this position, the input circuit to the amplifier is shorted to permit the electrodes to be removed from the solution without the drift caused by the open circuit. Whenever the switch is in this position, the meter will read 0 mV or pH 7.

Because the measuring electrode is at such a high impedance above ground, it is also quite susceptible to electrical interference. The electrode and its lead must be shielded (see Fig. 1.3). The shield is open inside the electrode, but connected to the reference-electrode lead within the measuring instrument. Or, as shown in Fig. 2.5, the shield may be driven by the transmitter amplifier so that it is at the same voltage as the measuring electrode but at a low impedance to ground. Stray signals are then not readily induced in the lead.

A driven shield does not present a time lag, as does a shield at ground or reference potential, since its capacitance does not charge or discharge with changes in measured voltage. With such a circuit the electrodes may be located as far as 2000 ft from the measuring instrument with negligible electrical interference and virtually instantaneous response.

Figure 2.5 shows ac paths to ground from the shield and reference lead through a capacitor. It is not customary for manufacturers to ground an instrument circuit directly, since input leads

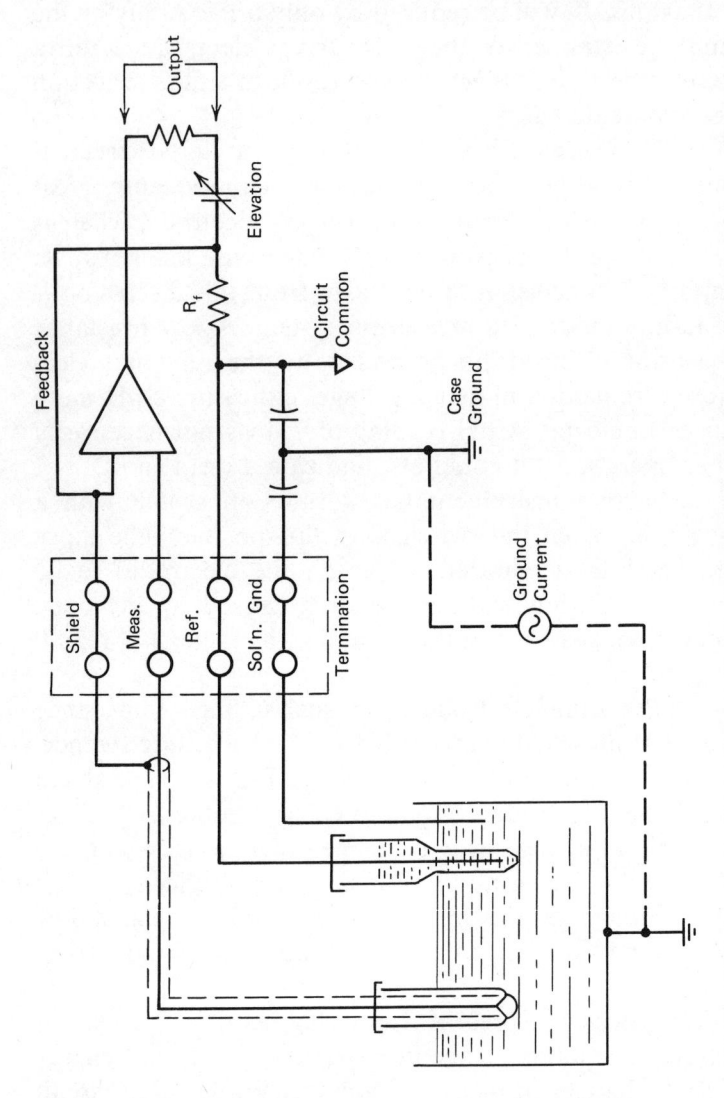

FIGURE 2.5. Proper shielding and grounding are both essential to satisfactory performance.

are always susceptible to grounding. Instead, common points in the circuit are typically grounded through capacitors.

Since the solution and the instrument ground are not necessarily at the same potential, a ground current could flow from one to the other through the reference electrode. This current may, in some cases, develop a significant potential across the liquid-junction resistance, causing an error. Even an ac current can cause an error, for the Ag, AgCl electrode acts as a rectifier. The solution-ground wire shown in Fig. 2.5 is connected from the electrode housing to instrument ground through a capacitor, effectively bypassing ac ground currents around the reference electrode, thereby minimizing this source of trouble.

Temperature Compensation

The Nernst equation indicates the variation of the pIon coefficient with absolute temperature [see (1.2)]. With every electrode pair, an isopotential point exists at which no temperature error exists. All millivolt-versus-pIon lines for various temperatures pass through that point. If the internal cell in the measuring electrode is identical to that of the reference electrode, the isopotential point is coincident with 0 mV. In the case where the measuring electrode has no internal Ag, AgCl half-cell, as with the silver sulfide electrode, the isopotential point is not zero, and temperature compensation is more demanding.

A plot of the relationship between pH and millivolts for the glass electrode at various temperatures is given in Fig. 2.6. Those systems which are ordinarily operated in the vicinity of pH 7, at temperatures near ambient—and this includes most waste-treatment systems—do not require temperature compensation. The error caused by operating between 40 and 80°F over the range from pH 6 to 8 without temperature compensation is only ±0.03 pH units. This is hardly a significant error. Should the pH deviate more markedly from 7 due to some upset, the error will be larger but may still be unimportant. If the pH indicated 3 instead of being at the control point of 7, it would not bother the operator if the pH were actually 3.2—he is much more concerned with the 4 pH deviation than the 0.2 pH error.

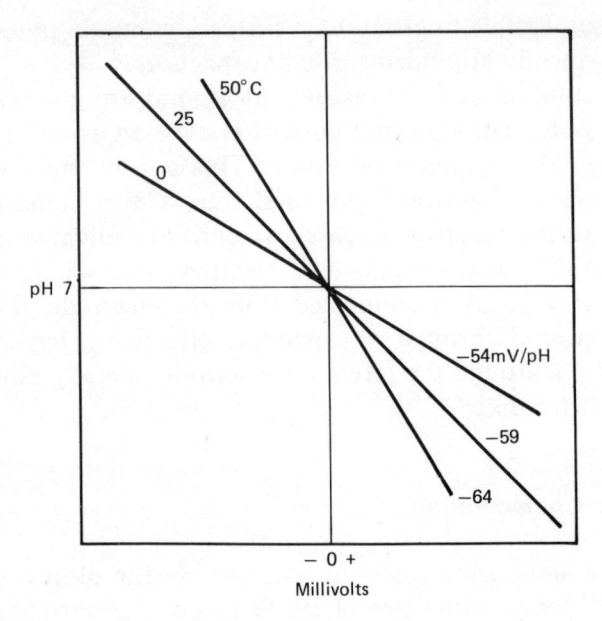

FIGURE 2.6. No temperature compensation is needed at pH 7.

If the process were to be normally operated at pH 3, however, an error of 0.2 pH units would be considerable, and temperature compensation should therefore be used. But perhaps because so few people understand the nature or significance of the temperature error in pH measurements, automatic compensation seems to be used much more often than it is really required.

Temperature compensation may be applied by adjusting the calibrated feedback resistor R_f, shown in Fig. 2.5. The output current from the transmitter passes through R_f, generating a millivoltage which the amplifier balances against the electrode potential. Reducing R_f by a token amount reduces the millivolt feedback for a given current output; but since the millivolt feedback is forced to match the electrode potential, the feedback does not increase, but instead the output current increases. In other words, a lower temperature, generating a lower cell potential at a given pH, is compensated by a lower value of R_f.

Making R_f adjustable by means of a dial calibrated in degrees Fahrenheit or Celsius allows manual temperature compensation, available in every pH measuring system. Automatic temperature compensation can be provided by making part of R_f a resistance-temperature bulb mounted in the electrode assembly. In instruments where automatic temperature compensation is provided, a switch is available to select manual compensation if desired for maintenance purposes.

All electrodes with internal Ag, AgCl cells are temperature compensated the same way as the pH electrode. The temperature characteristic of the fluoride-ion electrode, for example, is similar to that shown in Fig. 2.6, but with lines intersecting at 1 ppm F^- and 0 mV. Solid-state electrodes without internal Ag, AgCl cells do not have an isopotential point fixed by the filling solution because there is none. Consequently, the isopotential point is not then located at a convenient point in the operating range, making temperature compensation not only essential but more difficult.

As an example, the combination of an Ag_2S measuring electrode with a 1 M Ag, AgCl reference electrode has an isopotential point of -758 mV, corresponding to pS 4.6, as shown in Fig. 2.7. If the device is used as a silver electrode, the isopotential point is really the same, but the electrodes are connected in reversed polarity, so that it becomes $+758$ mV.

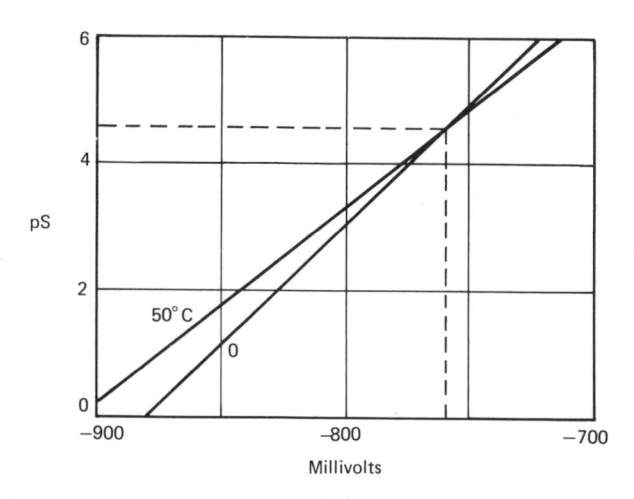

FIGURE 2.7. The Ag_2S electrode[3] has an isopotential point of -758 mV.

36 Measuring Systems

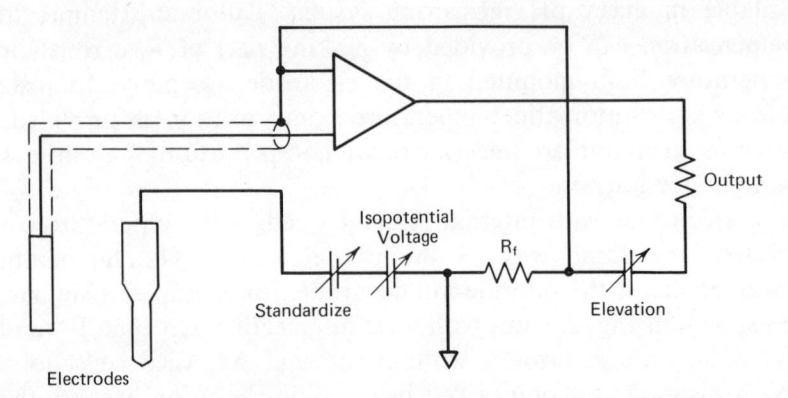

FIGURE 2.8. With unsymmetrical electrode combinations, the isopotential voltage must be introduced for effective temperature compensation.

In order to develop proper temperature compensation, which is a span correction about the isopotential point, the amplifier must be balanced about that point. Figure 2.8 shows how a bucking voltage set equal to the isopotential is applied to the input of the amplifier circuit. Since the 0-mV point is not likely to be at midscale output, an elevation potential must be applied at the amplifier output for proper positioning on the scale.

Adjustments

There are three general classes of ion transmitters—one for pH, a second for all other ion-selective electrodes, and a third for reduction-oxidation potential. Of the three, pH represents the highest volume of usage as well as the most standard application. The span of the pH transmitter should be adjustable from 2 to 14 pH units, over a range of -2 to $+16$ pH. Two calibration adjustments are ordinarily provided, one for each end of the scale. Manual or automatic temperature compensation may be selected by a switch, if the latter is required.

Finally, there is a standardization adjustment, a voltage applied to the input circuit, intended for canceling whatever asymmetry potential may be encountered. It is used to match the indicated pH against that of a buffer solution into which the electrodes are placed. Standardization may be switched out to determine how far out of calibration a given electrode pair would be without it. Switching the instrument to standby removes the standardization potential from the circuit, to avoid polarizing the electrodes when they are withdrawn from the solution.

Notice in Fig. 2.6 that increasing pH causes the electrode potential to decrease or to go further negative. To drive the output from the transmitter upscale then requires the input polarity to be reversed. This is the normal arrangement for pH instruments, and so does not require any special consideration. This point is important, however, with other electrodes which may be either cation- or anion-selective, and which may be calibrated in parts per million or pIon units (also opposite in sense). In the ion-selective transmitter, a bias adjustment of ±1000 mV is necessary to match asymmetry potentials of all the various electrodes. The span is adjustable typically between 50 and 1100 mV. Finally, a standardization adjustment is available for individual calibration of electrodes, as with the pH transmitter.

The reduction-oxidation potential transmitter does not ordinarily have temperature compensation or a standardization adjustment. As is seen in Chapter 5, interferences are many and spans are usually quite broad, so that high accuracy is usually not required. These instruments are normally calibrated directly in millivolts, with no polarity reversal available—increasing the oxidant activity always increases the electrode potential. Two-point calibration is ordinarily available over a range of −2000 to +2000 mV with a span adjustable between 100 and 1000 mV.

ELECTRODE ASSEMBLIES

As with many devices based on sound physical principles, failures often are due to the inadequacy of certain mechanical components. So it is with ion measurements—the electrode assembly is a very vulnerable part of the system. These devices have a very difficult task to accomplish. They must contain corrosive chemicals over a considerable range of temperatures and pressures while sealing rather delicate glass parts against leakage. The electrical side of the assemblies must be protected against leakage, too, while surrounded by all extremes of temperature, humidity, and precipitation. The assemblies must be easy to work on, and retain their seals notwithstanding frequent opening and closing. They are available in two basic designs—one for submerging in open vessels or streams, the second for piping under pressure.

The Submersible Assembly

In a typical submersible assembly (see Fig. 2.9), the electrodes are mounted through a casting supported by a vertical length of electrical conduit. They are sealed by rubber O-rings. A dome slips down the conduit over the casting and is threaded onto it with a large O-ring. Another O-ring seals the top of the dome against the conduit. The wires pass through the conduit and to a junction box where connection is made to a cable leading to the transmitter.

This cable must be flexible, to allow the assembly to be lifted from its position in the vessel or stream, for maintenance. It must also be sealed at its penetration into the junction box to keep out moisture.

To provide a constant flow of KCl electrolyte in those systems using a flowing reference electrode, the level of the solution must be higher than that in the process. This is assured by mounting a vented reservoir of about 250 ml capacity atop the junction box, with a tube connecting with the reference electrode through the conduit. At a typical flow of 1 ml/day, a filling should last 8 months.

Although the junction box is equipped with a gasketed cover, leaks manage to develop over a length of time. The most likely source of contamination is atmospheric. These assemblies are usually exposed to the weather, and diurnal temperature variations exert a considerable force on them. Overnight cooling forms a substantial vacuum in contrast to daytime temperatures in the open sun. Even a small leak can bring in a significant amount of humid air overnight with possible condensation. Expansion occurs during the heat of the day, but if condensation has developed within the assembly, water tends to collect at low points and not be exhausted through the leak. After many day-night cycles, and a few rain-sun

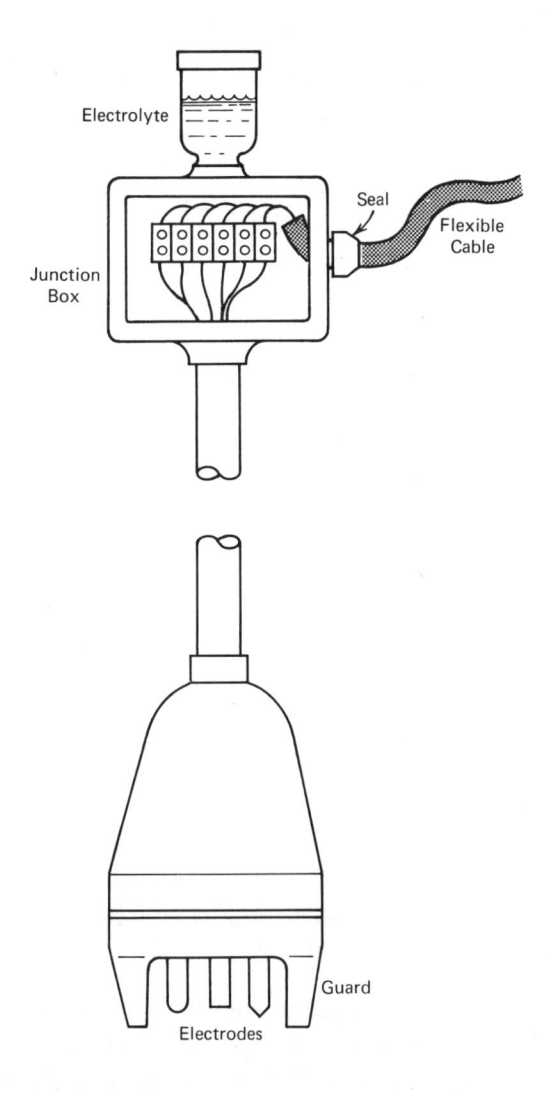

FIGURE 2.9. A Foxboro submersible assembly.

cycles, enough moisture can accumulate inside to coat terminal strips and cause a loss of signal.

Some engineers place packets of silica-gel or similar desiccant in the junction box to absorb this moisture. If there is a leak, however, the desiccant intensifies the vacuum by adsorbing moisture, thereby accelerating the accumulation until it is saturated. A wet desiccant

40   Measuring Systems

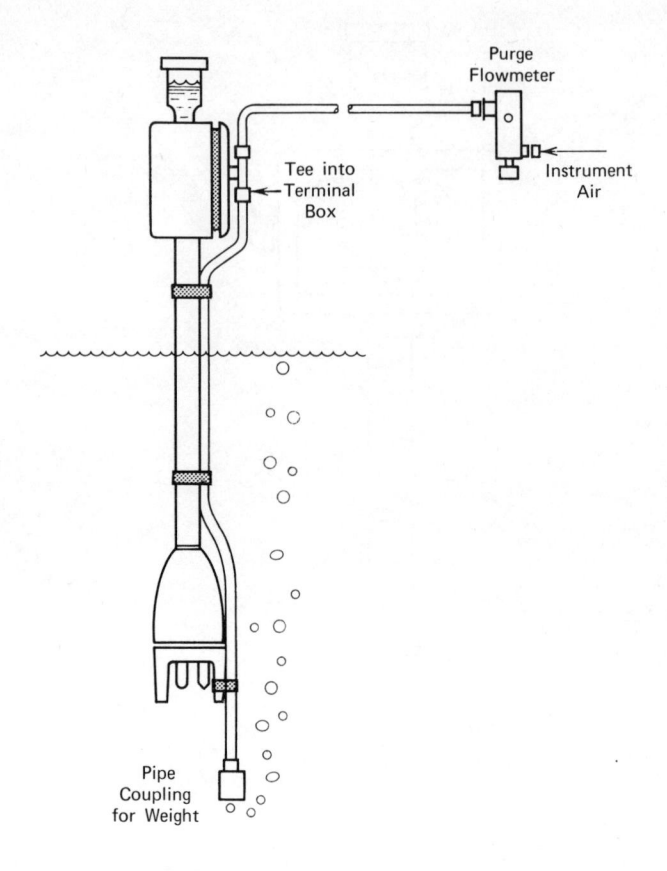

FIGURE 2.10. The bubbler maintains a constant differential pressure across the seals of the assembly, keeping moisture out.

packet resting against terminals can do more harm than a few isolated drops of water, so this is not a satisfactory remedy.

The best solution to this problem seems to be air-purging of the assembly. If the cover of the junction box is drilled and tapped for $\frac{1}{4}$-in. pipe, it may be connected to a source of instrument air as shown in Fig. 2.10. With the open end of the tube extending below the electrodes, the bubbling air will build up a backpressure inside the assembly just sufficient to prevent leakage from without. In event of a penetration, air will leak out, instead of moisture leaking in. Instrument air is usually dried to a dew point of about $-30°F$, so that condensation inside the electrode assembly is impossible.

Flow-through Assemblies

Although it would be possible to insert electrodes directly into a stream flowing in a pipe under pressure, maintenance prohibits such a practice. Electrodes are subject to failure and therefore must be accessible for replacement or cleaning without shutting down the process. Therefore the assembly which is designed for operation under pressure is not normally inserted directly in the process stream but in a side stream or circulating loop.

While measuring a continuous sample does simplify maintenance, the measurement is not necessarily representative of the true condition of the process. Every effort should be made to withdraw a representative sample and minimize the time required to transport it to the electrodes.

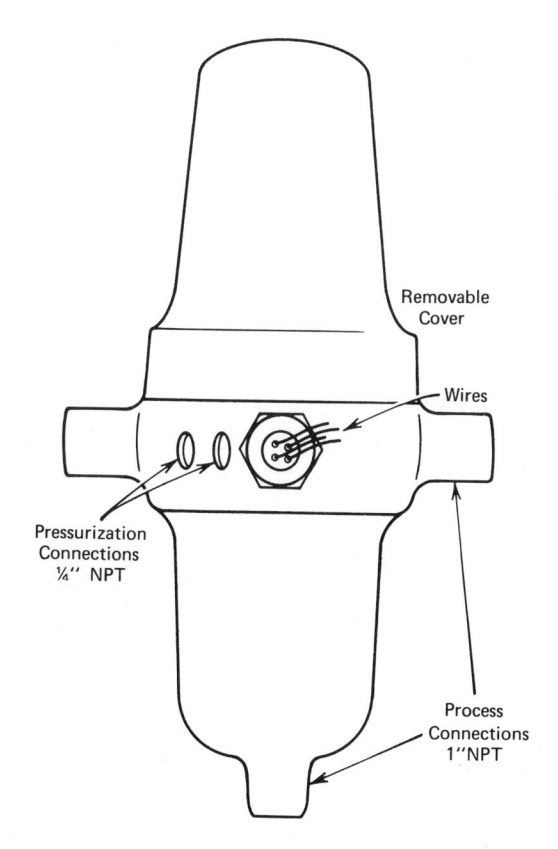

FIGURE 2.11. The Foxboro flow-through assembly.

42 Measuring Systems

The flow-through assembly is fitted with pipe connections for a continuous flowing sample at pressure up to 150 psig. Necessary whenever an ion measurement must be made under pressure or in a closed system, it is also preferred for batch processing in open vessels which are periodically emptied, which procedure would expose a submersible assembly.

Figure 2.11 shows how a typical flow-through cell is designed. The sample may pass through from one side to the other with the bottom connection used for draining, or may enter the bottom and overflow a side connection. Block valves should be installed on both sides of the assembly to allow the electrodes to be removed for maintenance.

When a flowing reference electrode is used with this type of assembly, provisions must be made for pressurizing it above the process. In cases where the sample may be sent to an open drain, the measurement may be made at atmospheric pressure, so that no special precautions need be taken. In some applications, however, the sample must be returned to the process under pressure, or in other cases release of pressure will change the composition of the sample by allowing gas to escape. Then the head space of the assembly should be pressurized with an inert gas to maintain a positive flow of electrolyte.

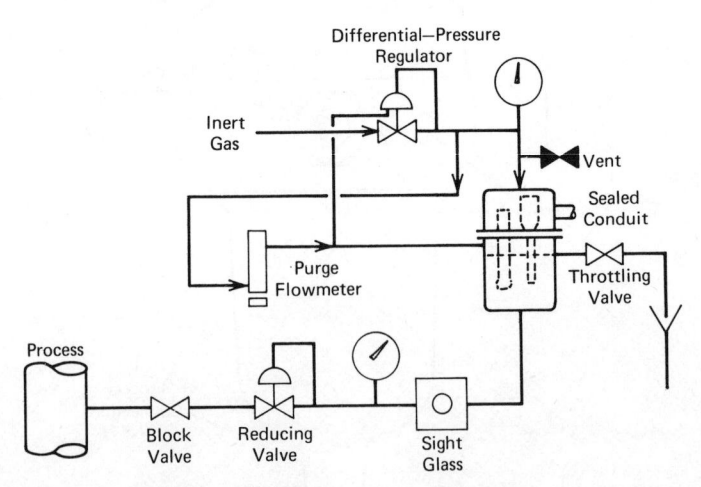

FIGURE 2.12. A pressure difference of 3 psi across the reference electrode assures constant flow of electrolyte.

Differential-pressure regulators are available which can hold a fixed 3-psi elevation above any process pressure up to 150 psig. Figure 2.12 is a schematic of a system used to measure the pH of a solution of a dissolved gas under pressure. The purge flowmeter is used to keep process fluid or corrosive gas from forcing its way into the regulator. The reducing valve should be set just high enough to keep the process gas in solution as indicated in the sight-glass.

MAINTENANCE

There are two completely separate calibrations to be made on an ion-measuring system. The transmitter must first be calibrated with a millivolt potentiometer to the range desired for the measurement. Then the electrode must be inserted and checked against a solution of known composition. Whatever calibration is done on them is termed standardization, that is, making them conform to whatever standard solution is used. The two procedures must be kept separate and distinct, principally because virtually every case of failure or inaccuracy is due to electrode or wiring problems—not the transmitter.

Transmitter Calibration

A transmitter calibration should be made at both ends of the scale using a portable potentiometer to introduce appropriate millivolt input signals, simulating the electrodes. If an unsymmetrical electrode combination is used such as the sulfide electrode with a silver-silver chloride reference, the transmitter must first be balanced at midscale output (0 mV) at the isopotential point.

Following a span adjustment, the entire range may then be elevated to desired operating level. Each style of transmitter will have its own procedure in this regard, so manufacturers' instructions should be closely followed.

Standard Solutions

Proper operation of the electrodes can only be demonstrated with reliability by checking them in a solution of known composition. A variety of commercial solutions are available for most of the applications commonly encountered. They have been especially compounded to be as stable as possible with time and temperature. Most are available as solutions, but some may now be purchased in dry form for easier transportation. Table 2.1 lists some important standard solutions used by The Foxboro Company, along with their formulation.

The pH standard solutions may properly be called "buffers" in that they contain a quantity of hydrogen not ionized at that particular pH level. As a result of this inventory, they resist changes in pH, even in the face of moderate dilution or contamination with other acids and bases. (Distilled water is so easily contaminated that it should never be used to check a pH electrode.)

The other solutions listed have no buffering action, and so are sensitive to dilution and contamination by interfering or reacting compounds. A stabilizing medium is added to some of these solutions as indicated to establish the proper environment for the calculated activity to be realized. The last solution on the list may be used to check any reduction-oxidation electrode assembly whose range covers the particular millivolt potential given.

Standardization

In theory, an electrode pair should develop a signal very close to that indicated for the given standard solution. All other things considered, only the small asymmetry potential need be adjusted with the standardizing knob. As electrodes age, this potential will gradually change, but only at the rate of a few millivolts per year.

Unfortunately, the standardizing adjustment is used to force an agreement between electrodes and standard solution even when they differ markedly. Large discrepancies—50 mV or more—should not be "covered up" in this way, but should be diagnosed. They are an indication of an impending failure or a rapidly deteriorating system.

Notice that two or more standard solutions are given for each ion. This is necessary to determine the source of any significant error. If, for example, a set of pH electrodes indicated pH 7 with a pH 6.86 buffer, and pH 7.5 with a pH 9.18 buffer, electrical leakage

Table 2.1 Standard Solutions[4]

Ion	Concentration	Activity	Composition
H$^+$	pH 4.008 6.86 9.18	pH 4.008 6.86 9.18	0.05 M KH phthalate 0.025 M KH$_2$PO$_4$ + 0.025 M Na$_2$HPO$_4$ 0.01 M borax
F$^-$	10.0 ppm F$^-$ 1.0	9.8 ppm F$^-$ 1.0	22.10 mg NaF/liter 2.21 mg NaF/liter
Cl$^-$	pCl 2.0 3.0	0.0061 M Cl$^-$ 0.00061	0.01 M KCl in 1 M KNO$_3$ 0.001 M KCl in 1 M KNO$_3$
Ag$^+$	pAg 2.0 3.0	0.009 M Ag$^+$ 0.0096	0.01 M AgNO$_3$ 0.001 M AgNO$_3$
S^{2-}	pS 1.0 3.0	0.015 M S^{2-} 0.00015	0.1 M Na$_2$S in 1 M NaOH 0.001 M Na$_2$S in 1 M NaOH
CN$^-$	pCN 3.0 4.0	0.76×10^{-3} M CN$^-$ 0.76×10^{-4}	10^{-3} M NaCN in 0.1 M NaOH 10^{-4} M NaCN in 0.1 M NaOH
Ca^{2+}	1000 ppm CaCO$_3$ 10	0.55×10^{-2} M Ca^{2+} 0.92×10^{-4}	10^{-2} M CaCl$_2$ 10^{-4} M CaCl$_2$
Redox[5]	0.1 M Fe^{2+} 0.1 M Fe^{3+}	+439 mV vs. 1 M Ag, AgCl	0.1 M Fe(NH$_4$)$_2$(SO$_4$)$_2$ 0.1 M FeNH$_4$(SO$_4$)$_2$ 1.0 M H$_2$SO$_4$

would be indicated which is reducing the sensitivity of the signal. On the other hand, if the electrodes indicated pH 8.5 and 10.8 in the same two buffers, it may be caused by contamination of the reference electrode.

More than one stream has been polluted by an operator who could not standardize his electrodes against two different buffers, so settled for one (and an insensitive electrode). And large elevation changes caused by contamination of the reference electrode are warnings of more problems to come—possibly imminent failure. Whereas manufacturers are typically generous with the adjustability of the standardization potentiometer, the amount really needed should be checked periodically by switching it out of the circuit and observing the resulting change. When the amount of standardization exceeds a few millivolts, it is time to look for trouble.

Observe the considerable difference between concentration and activity in some of the standard solutions. They have been prepared for concentration calibrations since that is what is usually requested. This is particularly important when comparison against a chemical analysis is required, as in water hardness. To avoid the concentration-activity argument, the user is urged to prepare his own standard solutions that will agree with whatever methods of analysis are used in his own plant.

Cleaning

External surfaces of the electrodes are sensitive to coating with crystalline or amorphous solids, oils, or greases.

Accuracy and dynamic response of all electrodes are severely degraded by these coatings. Response time which is normally a second or less has been observed to increase to several minutes even with only a thin (1 mm) coating of slime. The effect is similar to inserting a temperature bulb into a well of ever-increasing thickness. In a control loop, the presence of a substantial coating is indicated by periods of oscillation increased well beyond what is normally observed, for example from 2 min to half an hour or more.

The most effective cleaning procedure here depends on the nature of the fouling materials. Ultrasonic cleaners are available which vibrate solids loose from the electrodes. Ultrasonic cleaners are very effective against the buildup of crystalline solids like $CaSO_4$, muds, or thick syrups containing concentrated impurities from refining sugar. They are not effective against elastic coatings such as latex.

Where ultrasonic cleaners cannot be used, solvent cleaning may be practical. Certain solids are soluble in acids or bases—a solution of 10 to 20% H_2SO_4 is usually quite effective. Fat may be removed with hot water. Alcohols may be useful in other situations. When periodic washing with solvents is required, flow-through assemblies must be used so that an automatic solenoid valve may cycle the flow between wash and sample. If the coating is removed quickly, a single set of electrodes may be sufficient, with the controller, if any, transferred from automatic to manual during washing. For more adhesive coatings, two sets of electrodes may be necessary, with one in use while the other is immersed in the solvent. Switching the high-impedance leads from the electrodes is not recommended. Rather, two transmitters should be used, with switching on the output.

Internal parts of the electrode assembly may need cleaning from time to time, particularly if there has been an accumulation of moisture within it. This is most readily removed with methanol, acetone, or a similar volatile solvent which is miscible with water.

REFERENCES

1. R. G. Bates, *Determination of pH—Theory and Practice*, John Wiley & Sons, Inc., New York, 1964, p.303.
2. F. E. Moore, Taking Errors out of pH Measurement by Grounding and Shielding, *Instrument Society of America Journal*, February 1966.
3. L. E. Negus and T. S. Light, Temperature Coefficients and Their Compensation in Ion-Selective Measuring Systems, presented at the 18th National Symposium of Analytical Instruments Division of the Instrument Society of America, San Francisco, May 3–5, 1972.
4. "Ion-Selective Measuring Systems," *Technical Information Sheet* 43-31a, The Foxboro Company, Foxboro, Massachusetts.
5. T. S. Light, A Standard Solution for Oxidation-Reduction Potential (ORP) Measurements, *Analytical Chemistry*, May 1972.

PART 2

APPLICATION

3

THE HYDROGEN ION

A good pH electrode always tells the truth about the hydrogen-ion concentration in a stream. With media containing living organisms, like blood or river water, pH of itself is important. Often, however, we are not really interested in the hydrogen-ion concentration but rather in acidity or alkalinity, which are related to but not solely determined by hydrogen-ion concentration. This is particularly true in control work where acid must be balanced against base. The use of a pH measurement alone to infer acidity or alkalinity can be dangerous, but in most cases, this is the only measurement available. It is therefore essential that the relationships between pH and acid and base concentrations be firmly established before proceeding further.

STRONG ACIDS AND BASES

"Strong" reagents are defined as those which completely dissociate into their hydrogen or hydroxyl and companion ions, allowing their total acidity or alkalinity to be measured in terms of hydrogen-ion concentration. Weak reagents, by contrast, are only partially ionized, the balance of the hydrogen or hydroxyl groups remaining bound with the remainder of the molecule. The hydrogen-ion concentration of a weak acid is therefore only a small fraction of the total acid concentration. Next, however, let us define units of concentration.

Normality versus Weight Percent

Chemists generally work with acids and bases in units of normality, and this is the basis of the pH scale. Yet reagents used in the plant are customarily purchased or prepared on the basis of weight percent. One of the problems often encountered in plant design work is to determine how much reagent is needed to neutralize influent of a given pH or having a given titration curve. Titration curves are invariably determined in a laboratory using reagents whose concentrations are expressed in normality. So in either case, a conversion between normality and weight percent is required.

Weight percent is defined as the pounds (or grams) of full-strength chemical contained in 100 lb (or grams) of solution (or suspension). Normality is the gram-ions of replaceable hydrogen or hydroxyl groups in a liter of solution (or suspension).

To convert from weight percent to normality, it is first necessary to find the grams of reagent in a liter of solution using ρ, the density in grams per milliliter:

$$\text{wt. \%} = \text{g}/100 \text{ g solution}$$

$$100 \text{ ml} = 100 \text{ g}/\rho$$

Then

$$\text{g}/100 \text{ ml} = \rho(\text{wt. \%})$$

and

$$\text{g}/\text{l} = 10\rho(\text{wt. \%})$$

Next, divide by molecular weight M to obtain gram-moles per liter:

$$\text{g-mol}/\text{l} = \frac{10\rho(\text{wt. \%})}{M}$$

Finally multiply by the number n of replaceable hydrogen or hydroxyl groups per mole, to obtain normality:

$$N = \frac{10n\rho(\text{wt. \%})}{M} \qquad (3.1)$$

Example 3.1

Calculate the normality of a 10% (by weight) lime slurry whose density is 1.06 g/ml. The molecular weight of $Ca(OH)_2$ is 76 and $n = 2$.

$$N = \frac{10(10)(1.06)2}{76} = 2.78\ N$$

A list of normalities of some common reagents is given in Table 3.1.

Hydrogen Ions from Water

Water is a weakly dissociated molecule which is the solvent for most acids and bases:

$$H_2O \overset{K_w}{\rightleftarrows} H^+ + OH^-$$

We know that pure neutral water at 25°C has a pH of 7, indicating the extent of its ionization. An aqueous solution with a pH of 7 at

Table 3.1 Normalities of Common Reagents

Reagent	Wt. %	Normality (N)
HCl	32	10.17
	38	12.35
H_2SO_4	62.2 (50 Be′)	19.5
	77.7 (60 Be′)	27.2
	93.2 (66 Be′)	35.2
	98.0	36.0
NaOH	10	2.75
	20	7.93
	50	19.1
$Ca(OH)_2$	5	1.36
	10	2.78
	15	4.30

25°C does not contain a completely ionized acid of concentration 10^{-7} N—it is the water itself contributing the hydrogen ions. According to the definition of the equilibrium constant, K_w:

$$[H^+][OH^-] = K_w = 10^{-14} \quad \text{at } 25°C \qquad (3.2)$$

(The bracketed terms represent the concentration of the species within, in normality.) When $[H^+]=[OH^-]$, $[H^+] = \sqrt{K_w} = 10^{-7}$, defining neutrality as pH 7. It should be remembered that the neutral point is temperature dependent, however: $\log K_w$ varies from about -14.9 at 0°C to -13.3 at 50°C, and -12.1 at 100°C.[1]

To calculate the hydrogen-ion concentration of a given mixture of strong acid and base, this contribution from water must be taken into account, although it is only significant between pH 6 and 8. Consider the reaction between NaOH and HCl. Both are essentially completely ionized as indicated by their pH at the various concentrations given in Table 3.2. Their ionization and reaction will proceed as follows:

$$HCl + NaOH + H_2O \rightarrow H^+ + OH^- + Na^+ + Cl^-$$

The sum of all the negative and positive charges must be equal:

$$[Na^+] + [H^+] = [Cl^-] + [OH^-] \qquad (3.3)$$

Assuming complete ionization of both acid and base, $[Na^+]$ must equal the original concentration of base x_B, and $[Cl^-]$, the original concentration of acid x_A:

$$x_B + [H^+] = x_A + [OH^-]$$

Table 3.2[2] Measured pH versus Concentration for HCl and NaOH

Concentration (N)	pH HCl	pH NaOH
1.0	0.10	14.05
0.1	1.07	13.07
0.01	2.02	12.12

The relationship between [H⁺] and [OH⁻] has been established by (3.2), allowing substitution for [OH⁻]:

$$x_A - x_B = [H^+] - \frac{10^{-14}}{[H^+]} \qquad (3.4)$$

Equation (3.4) can then be used to relate acid and base concentration to pH:

$$x_A - x_B = 10^{-pH} - 10^{pH-14} \qquad (3.5)$$

Table 3.3 Theoretical Concentration of Strong Acids and Bases versus pH[a]

pH	x_A (N)	pH	x_A (N)
J.0	1.00×10^{-J}	6.0	0.99×10^{-6}
J.1	0.80	6.1	0.78
J.2	0.63	6.2	0.62
J.3	0.50	6.3	0.48
J.4	0.40	6.4	0.37
J.5	0.32	6.5	0.28
J.6	0.25	6.6	0.21
J.7	0.20	6.7	0.15
J.8	0.16	6.8	0.096
J.9	0.13	6.9	0.046
		7.0	0.000

pH	x_B (N)	pH	x_B (N)
J.0	$1.00 \times 10^{J-14}$	7.0	0.00
J.1	1.3	7.1	0.046×10^{-6}
J.2	1.6	7.2	0.096
J.3	2.0	7.3	0.15
J.4	2.5	7.4	0.21
J.5	3.2	7.5	0.28
J.6	4.0	7.6	0.37
J.7	5.0	7.7	0.48
J.8	6.3	7.8	0.62
J.9	8.0	7.9	0.78
		8.0	0.99

[a] Here J is the integer of pH below the value of interest, except 6 and 7.

56   The Hydrogen Ion

To assist in determining quickly the concentrations of strong acids and bases from a pH measurement, Table 3.3 has been prepared by substituting pH values into (3.5). First x_B was set to zero to determine x_A, then x_A was set to zero, to calculate x_B.

As an example of how the table is used, consider a solution at pH 12.4. Its base concentration would be $2.5 \times 10^{12-14}$ or 2.5×10^{-2} N. The contribution of water is seen by comparing the right- and left-hand columns. At pH 6 or 8, water contributes only 1% to the hydrogen- or hydroxyl-ion concentration, but at pH 7, it contributes 100%.

Table 3.3 actually tabulates the titration curve of completely ionized agents. Its slope at any point may be determined by differentiating (3.4):

$$\frac{d(x_A - x_B)}{d[H^+]} = 1 + \frac{10^{-14}}{[H^+]^2}$$

Next, the relationship between pH and $[H^+]$ must be differentiated:

$$pH = -\log[H^+]$$

$$\frac{dpH}{d[H^+]} = -\frac{\log e}{[H^+]} = -\frac{0.434}{[H^+]}$$

Then the two derivatives are combined:

$$\frac{dpH}{d(x_A - x_B)} = \frac{dpH/d[H^+]}{d(x_A - x_B)/d[H^+]}$$

$$\frac{dpH}{d(x_A - x_B)} = -\frac{0.434}{[H^+] + 10^{-14}/[H^+]}$$

$$\frac{dpH}{d(x_A - x_B)} = -\frac{0.434}{10^{-pH} + 10^{pH-14}} \qquad (3.6)$$

The maximum slope occurs at neutrality, that is, at pH 7:

$$\frac{dpH}{d(x_A - x_B)} = -\frac{0.434}{10^{-7} + 10^{-7}} = -2.17 \times 10^6 \, pH/N \qquad (3.7)$$

At every other pH, the slope will be less, as one of the terms in the denominator grows while the other diminishes.

Other strong acids include HBr, HI, $HClO_4$, and HNO_3. Although some consider H_2SO_4 as a strong acid, it is not completely ionized below pH 2, and its pH versus concentration profile in that region will deviate from that of HCl. Strong bases include LiOH and KOH. Here again, $Ca(OH)_2$ is not completely ionized above pH 12 and will be treated under weak bases.

The pH scale is not bound to the limits of 0 to 14. Concentrations of strong acids or bases exceeding 1.0 N will register pH values below 0 and above 14. In these regions, nonidealities begin to dominate, particularly variations in the activity coefficient of both the reagent and water itself, which is normally assumed to be unity. A concentrated solution of $HClO_4$, for example, can register a pH below its equivalent concentration, having drastically reduced the activity of the water by capturing a large portion of it for hydration. Nonetheless, measurements can be made in these regions if calibration is made specifically for the materials to be measured.

WEAK ACIDS AND BASES

Weak reagents are only partially ionized such that their actual concentrations are always greater than the respective hydrogen- or hydroxyl-ion concentration. Thus they invariably will require more reagent to neutralize them than their pH would indicate. Exact equivalence of acid and base does not necessarily correspond to pH 7 as with strong agents and, in fact, varies with the concentration of the salt produced. The slope of the titration curve is much more gradual than for strong agents and is influenced by the presence of salts. Add to this the contribution of a second or third hydrogen or hydroxyl ion and the relationship between concentration and pH becomes quite complex.

Ionization of Weak Reagents

To demonstrate the properties of a weak agent, let us add some acetic acid (abbreviated HAc) to the mixture of HCl and NaOH previously investigated:

$$HCl + NaOH + HAc + H_2O \rightleftharpoons H^+ + OH^- + Na^+ + Cl^- + Ac^-$$

58 The Hydrogen Ion

Only a fraction of weak acid will ionize, the remainder being in equilibrium with the product of its ions, obeying the relationship:

$$K_C = \frac{[H^+][Ac^-]}{[HAc]} \qquad (3.8)$$

A charge balance on the system is as follows:

$$[H^+] + [Na^+] = [OH^-] + [Cl^-] + [Ac^-]$$

Here again, x_B and x_A may be substituted for $[Na^+]$ and $[Cl^-]$, due to complete ionization. The initial concentration of the third reagent, acetic acid, is given as x_C, and is the sum of the undissociated acid and the acetate ion:

$$x_C = [HAc] + [Ac^-] \qquad (3.9)$$

The concentration of undissociated HAc may be found from (3.8):

$$[HAc] = \frac{[H^+][Ac^-]}{K_C}$$

Then,

$$x_C = \frac{[H^+][Ac^-]}{K_C} + [Ac^-]$$

Solving for $[Ac^-]$,

$$[Ac^-] = \frac{x_C}{1 + [H^+]/K_C}$$

Making all the appropriate substitutions into the ion balance yields

$$x_A - x_B = [H^+] - \frac{10^{-14}}{[H^+]} - \frac{x_C}{1 + [H^+]/K_C} \qquad (3.10)$$

Equation (3.10) can be applied to any system containing a monoprotic* weak acid alone, or titrated with a strong base or a strong

*Containing a single ionizable hydrogen atom per molecule.

acid. For convenience, (3.10) can be restated in terms of pH

$$x_A - x_B = 10^{-pH} - 10^{pH-14} - \frac{x_C}{1 + 10^{pK_c - pH}} \qquad (3.11)$$

where pK_C (like pH) is the negative base-10 logarithm of the ionization constant.

For the situation where $x_A - x_B = 0$, pH is uniquely determined by x_C. However, this only applies to pure weak acid in solution. Titration with another agent introduces x_A or x_B, altering the relationship. For the pure component then, (3.11) reduces to

$$x_C = (1 + 10^{pK_c - pH})(10^{-pH} - 10^{pH-14}) \qquad (3.12)$$

At a pH equal to the pK_C, the weak acid is 50% dissociated (providing that the pH is below 6 where ionization of water starts to interfere).

Example 3.2

Calculate the concentration of acetic acid in water solution at pH values of 6, 4, and 2. Its pK is 4.75:

at pH 6, $\quad x_C = (1 + 10^{-1.25})(0.99 \times 10^{-6}) = 1.04 \times 10^{-6}$ N

at pH 4, $\quad x_C = (1 + 10^{0.75})(10^{-4}) = 6.62 \times 10^{-4}$ N

at pH 2, $\quad x_C = (1 + 10^{2.75})(10^{-2}) = 5.62$ N

Next consider the monoprotic weak base, aqueous ammonia, ionizing to satisfy the relationship:

$$K_D = \frac{[NH_4^+][OH^-]}{[NH_3]}$$

(Butler[3] disputes the existence of un-ionized ammonium hydroxide, representing aqueous ammonia by NH_3.) If we define the concentration of total ammonia as x_D, then

$$x_D = [NH_3] + [NH_4^+]$$

60 The Hydrogen Ion

Using the same procedure as established for the weak acid,

$$[NH_4^+] = \frac{x_D}{1 + [OH^-]/K_D}$$

$$x_A - x_B = [H^+] - \frac{10^{-14}}{[H^+]} + \frac{x_D}{1 + 10^{-14}/[H^+]K_D}$$

Finally,

$$x_A - x_B = 10^{-pH} - 10^{pH-14} + \frac{x_D}{1 + 10^{pK_D + pH - 14}} \quad (3.13)$$

Again, in absence of x_A and x_B,

$$x_D = (1 + 10^{pK_D + pH - 14})(10^{pH - 14} - 10^{-pH}) \quad (3.14)$$

Example 3.3

Determine the ammonia concentration for a solution at pH 10. Its pK is 4.75:

$$x_D = (1 + 10^{0.75})(10^{-4}) = 6.62 \times 10^{-4} \, N$$

Neutralization of Weak Acids and Bases

Having established the pH versus concentration relationship for pure weak acids and bases, the next subject is an examination of their titration curves. Acetic acid at pH 4 does not require the same amount of caustic to neutralize it as acetic acid whose pH has been raised from 2.5 to 4 with caustic. The difference lies in the amount of acetate ions present, which exert a powerful influence over the ionization of the acid.

For neutralization of a weak acid with a strong base, only the applicable portion of (3.11) is used:

$$x_B = \frac{x_C}{1 + 10^{pK_C - pH}} - 10^{-pH} + 10^{pH - 14} \quad (3.15)$$

Here as in other formulas in this chapter, values are selected for x_C

Weak Acids and Bases 61

and pH, and x_B is calculated. Figure 3.1 compares the titration of 0.01 N HAc and 0.01 HCl with NaOH.

Example 3.4

Calculate the concentration of caustic required to raise the pH of an acetic acid solution from 4 to 6. Using Example 3.1, x_C at pH 4 is 6.62×10^{-4} N:

$$x_B = \frac{6.62 \times 10^{-4}}{1 + 10^{-1.25}} - 0.99 \times 10^{-6} = 6.36 \times 10^{-4} \ N$$

A weak base neutralized by a strong acid will be described by the appropriate terms of (3.13):

$$x_A = \frac{x_D}{1 + 10^{pK_D + pH - 14}} + 10^{-pH} - 10^{pH - 14} \qquad (3.16)$$

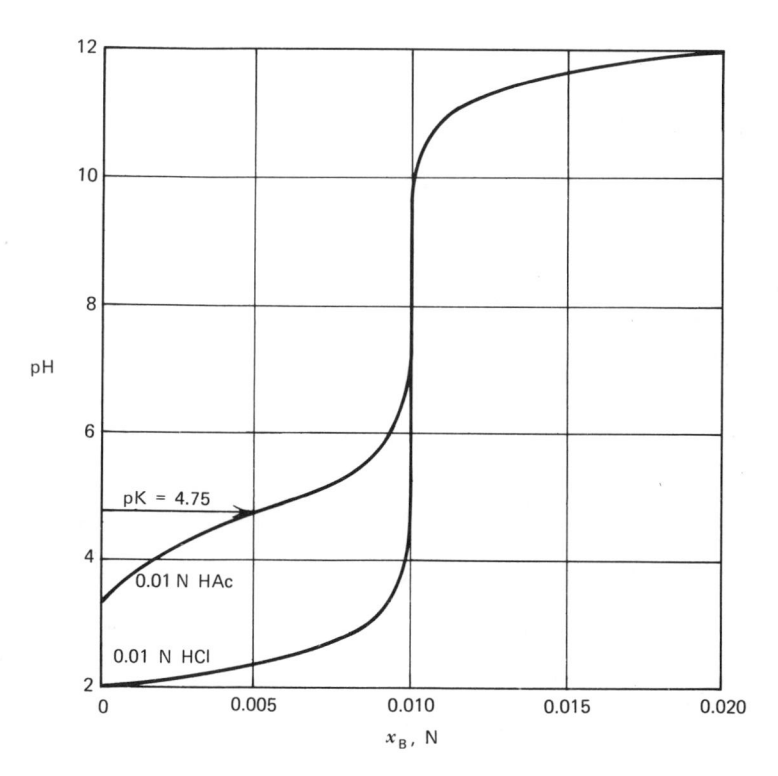

FIGURE 3.1. Comparison between titration curves of a strong acid and a typical weak acid.

62   The Hydrogen Ion

Here again, a titration curve can be generated by finding the acid required to lower the pH of a given concentration x_D to various selected values.

The equivalence point is not at pH 7 when weak acids or bases are neutralized. To find this point, set $x_B = x_C$ in (3.15). Then,

$$x_C = (10^{pH-14} - 10^{-pH})(1 + 10^{pH-pK_C}) \tag{3.17}$$

The solution appears to require trial and error. However, in many examples, the equivalence point will exceed pH 8, such that the terms 10^{-pH} and 1 in (3.17) become insignificant. The equation then reduces to

$$x_C \cong 10^{pH-14+pH-pK_C}$$

and

$$pH \cong \tfrac{1}{2}(14 + pK_C + \log x_C) \tag{3.18}$$

Example 3.5

Calculate the pH of a 0.01 N solution of sodium acetate:

$$pH \cong \tfrac{1}{2}(14 + 4.75 - 2) = 8.375$$

Check the solution against (3.17):

$$x_C = (10^{-5.625} - 10^{-8.375})(1 + 10^{3.625})$$

$$x_C = (2.37 \times 10^{-6} - 4.22 \times 10^{-9})(4.22 \times 10^3 + 1)$$

$$= 0.01\ N$$

The equivalence point for a weak base is found in like manner, setting $x_A = x_D$:

$$x_D = (10^{-pH} - 10^{pH-14})(1 + 10^{14-pH-pK_D}) \tag{3.19}$$

And the approximation when pH < 6,

$$pH = \tfrac{1}{2}(pK_D - 14 - \log x_D) \tag{3.20}$$

Buffering

Buffering is defined as the ability of a solution to resist changes in pH. All weak agents can buffer by converting hydrogen or hydroxyl ions into undissociated acid or base. To do this effectively, however, there must be a quantity of un-ionized agent available, and a source of companion ions. For example, acetic acid has no buffering effect above pH 6 where it is almost completely ionized. At low pH values, it has a great capacity to absorb hydroxyl ions due to its reserve of undissociated acid. But to absorb hydrogen ions with equal effectiveness, a supply of acetate ions must also be available.

Standard buffer solutions are therefore made from mixtures of weak acids and their strong-base salts, or weak bases and their strong-acid salts. They are most effective when both species are present in equal amounts, at which point, pH$\cong$pK_C or $14-$pK_D. The exact pH of such a mixture can be found by substituting x_B for the concentration of the salt and $x_C - x_B$ for the concentration of the acid, and by following (3.15). For a basic buffer, substitute x_A for the salt and $x_D - x_A$ for the base, and use (3.16).

Having the capacity to absorb hydrogen and hydroxyl ions makes the pH of solutions of weak agents much easier to control than those of strong agents. To demonstrate this point, the slope of a weak-acid titration curve will be found by differentiating (3.10):

$$\frac{dx_B}{d[H^+]} = -1 - \frac{10^{-14}}{[H^+]^2} - \frac{x_C/K_C}{(1+[H^+]/K_C)^2}$$

Following the same procedure used to generate (3.6),

$$\frac{dpH}{dx_B} = \frac{0.434}{[H^+]+10^{-14}/[H^+]+([H^+]x_C/K_C)/(1+[H^+]/K_C)^2}$$

Converting to pH units,

$$\frac{dpH}{dx_B} = \frac{0.434}{10^{-pH}+10^{pH-14}+\left[x_C 10^{pK_C-pH}/(1+10^{pK_C-pH})^2\right]} \quad (3.21)$$

Example 3.6

Calculate the slope of the titration curve at pH 7 for 0.01 N acetic acid neutralized with caustic:

$$\frac{d\text{pH}}{dx_B} = \frac{0.434}{10^{-7}+10^{-7}+(0.01\times 10^{-2.25})/(1+10^{-2.25})^2}$$

$$\frac{d\text{pH}}{dx_B} = \frac{0.434}{2\times 10^{-7}+(0.562\times 10^{-4})/(1.0056)^2}$$

$$= 0.78\times 10^4 \text{ pH}/N$$

The example indicates that 0.01 N acetic acid will give a titration curve over 200 times less sensitive at pH 7 than a strong acid titrated against the same base. The dominant term in (3.21) is the product $x_C 10^{pK_C-\text{pH}}$. An increase in either x_C or pK_C will markedly decrease the slope of the curve at a given pH.

The slope of the weak-base titration curve may be found in like manner:

$$\frac{d\text{pH}}{dx_A} = \frac{-0.434}{10^{-\text{pH}}+10^{\text{pH}-14}+(x_D 10^{pK_D+\text{pH}-14})/(1+10^{pK_D+\text{pH}-14})^2} \tag{3.22}$$

Since the slope of these titration curves varies with the initial concentration of weak acid or base, there is no such thing as, for example, an "acetic acid titration curve." Each value of initial concentration will give a differently shaped curve, and a different sensitivity at pH 7. In fact, the slope varies almost exactly with the inverse of initial concentration, hence with the flow of reagent required for neutralization—a point that will become important when considering feedback control systems in Chapter 6.

Mixtures of Strong and Weak Acids

This combination is not as puzzling as it might at first seem. Actually the formulation of the pH versus concentration equation (3.10) developed earlier has all the necessary ingredients. As a strong acid is added to a weak one, the pH is lowered but not markedly. When weak acid is added to a strong one, the pH may be lowered slightly or not at all, depending on the value of $pK_C - pH$.

As a supplement to the curve given in Fig. 3.1, a titration curve of a mixture of 0.01 N HCl and 0.01 N HAc against NaOH is shown in Fig. 3.2. It starts before the point (pH 3.28) where the former curve begins. This simply illustrates that a strong acid can be construed mathematically as a negative strong base and vice versa. Mixtures of strong and weak agents will become important in the application of feedforward controls in Chapter 9.

"Strong" Weak Acids

Two acids that are generally considered strong are sulfuric and hydrofluoric. The first hydrogen ionizes from sulfuric completely

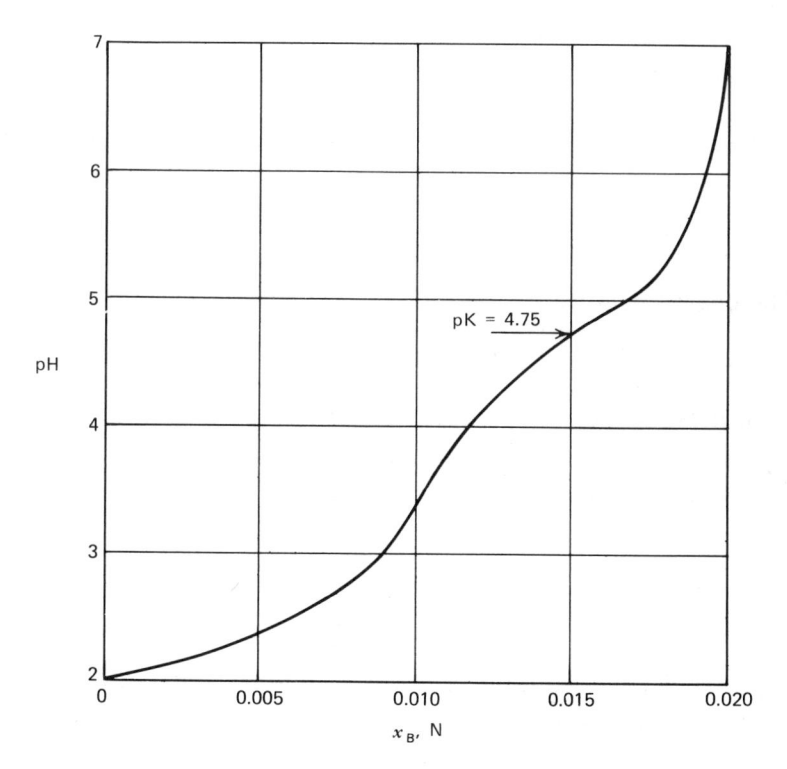

FIGURE 3.2. Titration of a mixture of 0.01 HCl and 0.01 HAc with NaOH.

but the second does not, so the acid behaves like an equal mixture of a strong and weak acid. Actually the second hydrogen is readily removed, as indicated by its pK of 1.99. For example, at pH 2, acid concentration is $0.0133N$. Perhaps the powerful dehydrating characteristic of sulfuric acid gives it a reputation as a strong acid.

Hydrofluoric acid (HF) is a weak acid whose pK is 3.17. Its ions do not etch glass, but rather the undissociated HF itself does. Above pH 3, etching is not a problem, but fluoride-bearing streams below pH 3 will etch. Glass electrode assemblies are particularly susceptible to attack.

POLYPROTIC SYSTEMS

Acids and bases with more than one hydrogen or hydroxyl ion to give up have more complex titration curves, and are generally more effective buffers than their monoprotic counterparts.

Diprotic Acids and Bases

Consider the reaction between the diprotic weak acid hydrogen sulfide and caustic:

$$NaOH + H_2S + H_2O \rightleftharpoons Na^+ + HS^- + S^{2-} + H^+ + OH^-$$

The charge balance is

$$[Na^+] + [H^+] = [HS^-] + 2[S^{2-}] + [OH^-]$$

Ionization of the acid proceeds as follows:

$$K_1 = \frac{[H^+][HS^-]}{H_2S}$$

$$K_2 = \frac{[H^+][S^{2-}]}{[HS^-]}$$

Total acid concentration in normal units is double the number of sulfur atoms present in their various states:

$$x_C = 2[H_2S] + 2[HS^-] + 2[S^{2-}]$$

Converting all terms to sulfide ion gives

$$x_C = 2[S^{2-}]\left\{1 + \frac{[H^+]}{K_2}\left(1 + \frac{[H^+]}{K_1}\right)\right\}$$

Substituting for [HS$^-$] and [Na$^+$] in the charge balance yields

$$x_B + [H^+] = [S^{2-}]\left(2 + \frac{[H^+]}{K_2}\right) + [OH^-]$$

Substituting for [S^{2-}] in terms of x_C gives

$$x_B + [H^+] = \frac{x_C(2 + [H^+]/K_2)}{2\{1 + ([H^+]/K_2)(1 + [H^+]/K_1)\}} + [OH^-]$$

Finally, converting [H$^+$] and [OH$^-$] to pH,

$$x_B = -10^{-pH} + 10^{pH-14} + \frac{x_C(1 + 0.5 \times 10^{pK_2-pH})}{1 + 10^{pK_2-pH}(1 + 10^{pK_1-pH})} \quad (3.23)$$

(A strong acid x_A could be simply added as was done with the derivation for monoprotic weak acids, but was here omitted for simplicity.)

Certain terms in (3.23) can be eliminated as insignificant, depending on the pH range evaluated. At pH values more than one unit above pK_1, the last term in the denominator becomes essentially unity; at pH values more than one unit below pK_2, the ratio of the terms containing pK_2 approaches 0.5. At pH = pK_1, half of the first ion group is neutralized, and at pK_2 half of the second group or three-fourths of the total is neutralized.

An application requiring the pH adjustment of a sulfide solution appears commonly in oil refineries. Sulfur is removed from various petroleum fractions along with nitrogen, and is collected as a slightly basic solution of ammonium sulfide known as "sour water." The sulfide must be stripped from the water as hydrogen sulfide gas, in which form it may be oxidized to elemental sulfur. But first the sulfide must be converted to H$_2$S by addition of sulfuric acid. If too little acid is added, sulfide will remain in solution and go unrecovered; if too much acid is added, it will be wasted and also promote severe corrosion. The problem is to control the pH at the optimum

68 The Hydrogen Ion

point for H$_2$S removal. The flowsheet is given in Fig. 3.3.

Modifying (3.23) by using acid concentration x_A in place of $-x_B$, and eliminating those terms which are insignificant at the low pH levels that will be encountered, develops

$$x_A = 10^{-pH} - \frac{0.5 x_C}{1 + 10^{pK_1 - pH}} \quad (3.24)$$

Here x_A will appear as the excess acid, the objective being to balance stoichiometrically whatever base is in solution to free the H$_2$S. For H$_2$S, pK_1 is 7.0 and pK_2 is 12.9; the latter is of no importance in this example.

Figure 3.4 gives curves of pH versus excess acid for three decades of sulfide concentration. At zero excess acid, we have the equivalence point typically obtained when neutralizing a weak base with a strong acid. In this example, the weak base is ammonium sulfide.

The concentration of 0.2 N H$_2$S represents nearly saturated solution at normal temperature and pressure. At the process temperature of 150°F, at least 35 psig backpressure must be maintained on the pH cell to keep the H$_2$S in solution. However, the concentra-

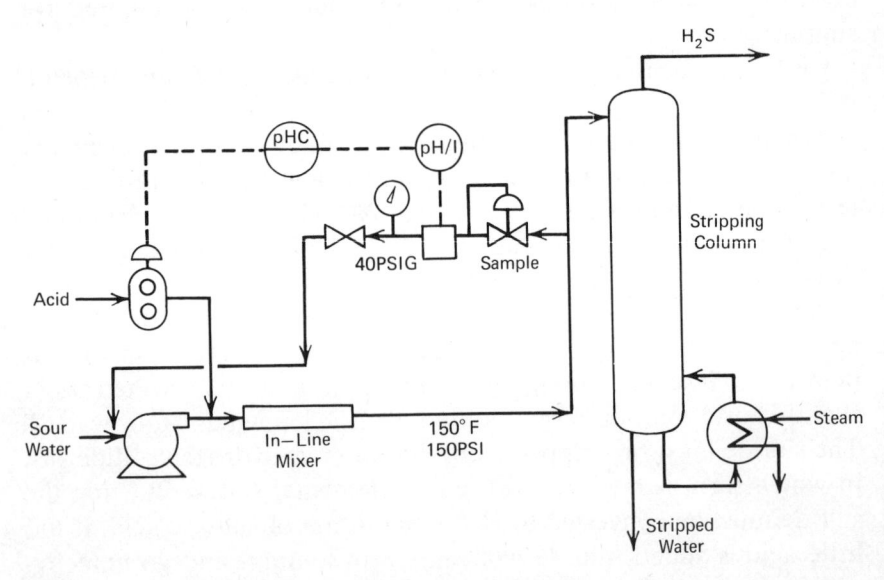

FIGURE 3.3. The flow of sulfuric acid is metered by a positive displacement pump to control the pH of the sour water.

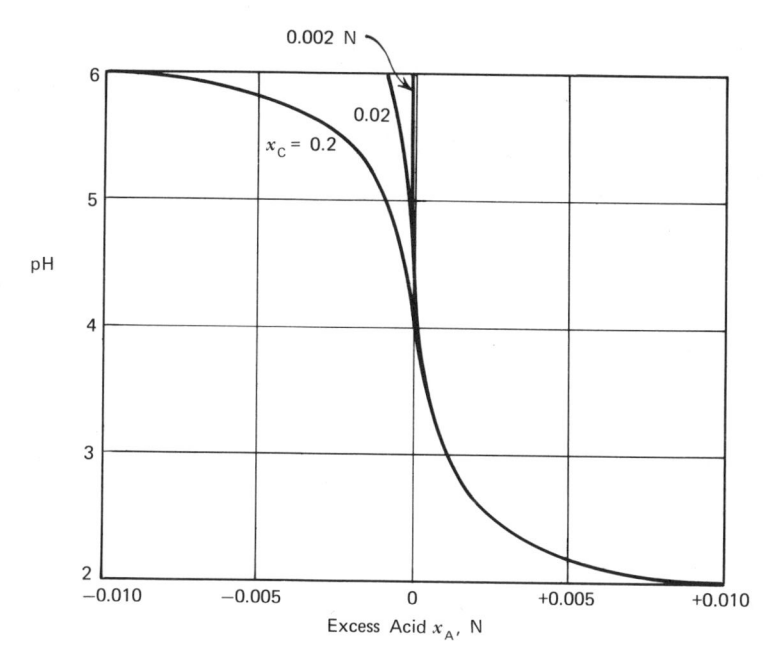

FIGURE 3.4. The pH corresponding to the concentration of excess sulfuric acid changes with sulfide concentration, x_C.

tion of H_2S is still about 0.2 N since the elevated temperature essentially cancels the effect of the elevated pressure on solubility. If the pH of the mixture is controlled at 4 by adding acid, it will rise to 5 or 6 following stripping. If the pH entering the stripping column is as low as 3, however, the stripped effluent will remain at pH 3.

Salts as Acids and Bases

Salts of strong bases and weak acids act as bases whose strength is in proportion to the difference in strengths of the constituents. Notable examples of basic salts are carbonates and bicarbonates. Alkali carbonates are quite basic and the bicarbonates slightly so. Since they appear in most groundwater supplies, their contribution to pH is worth examining.

Figure 3.5 is a titration curve of a 0.01 N solution of CO_2 with caustic, as developed by (3.23). At $x_B = x_C$ carbon dioxide and

70 The Hydrogen Ion

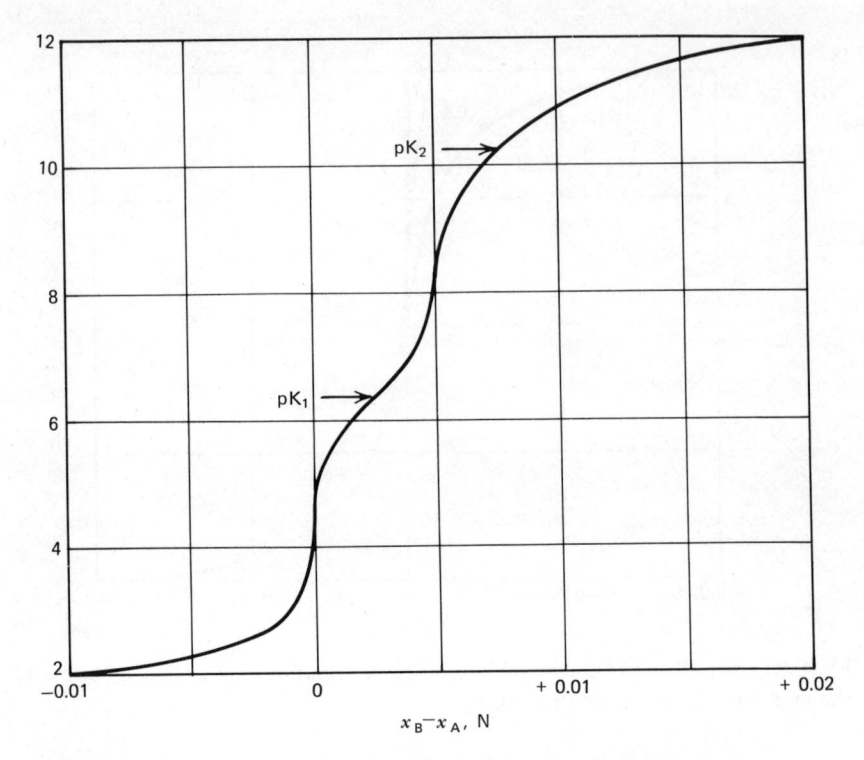

FIGURE 3.5. The titration curve for a 0.01 N carbonate solution.

sodium are in equivalence, yielding a 0.01 N solution of sodium carbonate. At $x_B = 0.5 x_C$, half the equivalent sodium is present, producing a 0.01 N bicarbonate solution. The value of pK_1 is 6.35 and that of pK_2 is 10.25; at these pH levels, $x_B = 0.25 x_C$ and $0.75 x_C$, respectively.

The most significant property of this curve is the gentle slope in the region of pH 6 to 7. Caustic solutions readily absorb carbon dioxide from the atmosphere, which considerably moderates their pH. Carbonate buffers therefore provide important regulation of pH in many aquatic systems.

Recognize that (3.7) represents the maximum possible slope of any aqueous pH titration curve. In fact the value of 2.17×10^6 pH/N can only be achieved at pH 7 in distilled water. Most processes use well, lake, or river water which contain a certain amount of bicarbonates, identifiable as alkalinity. *Total alkalinity* is reported as the amount of acid (in equivalent ppm $CaCO_3$) required to lower the pH of a water sample to 4.5. Since most water supplies have a pH less than 8, virtually all the alkalinity exists as bicarbonate.

To illustrate the buffering effect of alkalinity, consider a water source containing bicarbonate reported as 10 ppm $CaCO_3$—this is probably as high as normally encountered. Parts per million are the same as milligrams per liter; dividing by the molecular weight of $CaCO_3$ gives

$$\frac{10 \times 10^{-3} \text{ g/l}}{100 \text{ g/mol}} = 10^{-4} \, M \text{ CaCO}_3$$

Carbonate normality is twice the molarity above, and bicarbonate normality is four times as great, since

$$CaCO_3 + H_2O + CO_2 \rightarrow Ca(HCO_3)_2$$

Then,

$$10^{-4} \, M \text{ CaCO}_3 = 4 \times 10^{-4} \, N \text{ CO}_3^{2-} \text{ as HCO}_3^-$$

To determine the effect of this level of bicarbonate on the gain of the titration curve, (3.23) was differentiated and its slope evaluated at pH 7. Substituting 4×10^{-4} N for x_C with pK_1 of 6.35 and pK_2 of 10.25 yields

$$\frac{d\text{pH}}{dx_B} = 1.4 \times 10^4 \text{ pH}/N$$

The pH is over 100 times less sensitive to reagent addition than it is in distilled water. An alkalinity of 1 ppm $CaCO_3$ produces a slope at pH 7 of 1.3×10^5 pH/N, with intermediate values exhibiting a proportional slope.

Salts of various metals act as weak acids which are neutralized by conversion to their insoluble hydroxides. This is one of the common applications in metal-plating plants—removal of iron, chromium, zinc, and so on, from waste streams by precipitation. These ions exist only in acid solution. At pH 2, for example, all of the trivalent chromium in a solution is ionized. But as the pH is raised by a base, hydroxyl groups are successively added until at pH 8.5, a minimum

72 The Hydrogen Ion

solubility is reached. As more base is added, the chromium begins to redissolve as basic salts called chromites; zinc and aluminum also display this property, forming zincates and aluminates. A solubility curve for trivalent chromium[4] is given in Fig. 3.6.

Both ferric and ferrous iron are precipitated as hydroxides in similar fashion, although without redissolving in excess caustic. Again, hydroxyl groups are added stagewise, although in a typical titration curve[5] shown in Fig. 3.7, the individual steps are not distinguishable. Nonetheless, the presence of these metal ions contributes considerable buffering to the system.

The acidic hydrolysis of the ferrous ion, for example, proceeds as follows:

$$Fe^{2+} + H_2O \rightleftharpoons FeOH^+ + H^+$$

The pK for this equilibrium is given as 6.8, which fits the upper curve of Fig. 3.7 reasonably well. Addition of the second hydroxyl group by further titration with caustic causes precipitation above pH 8. By contrast, the ferric ion is a stronger acid, with pK values

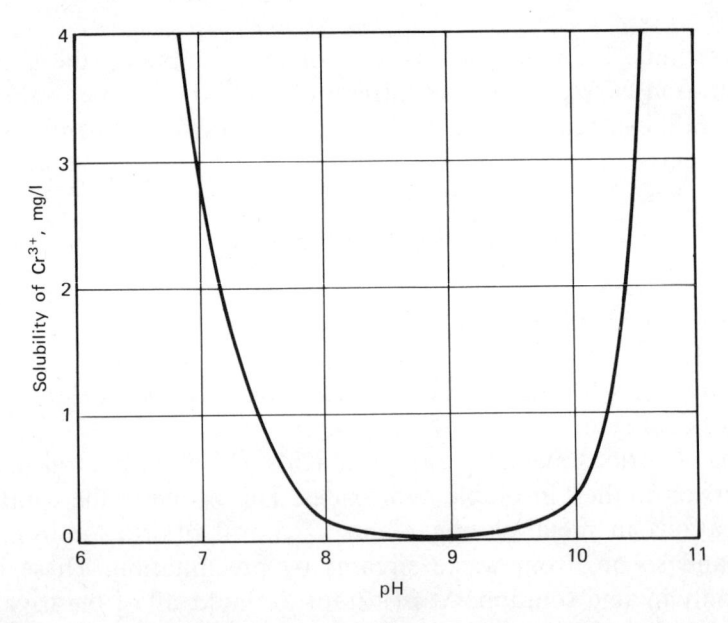

FIGURE 3.6. Chromium, like zinc and aluminum, can be precipitated as insoluble hydroxide by proper adjustment of pH.

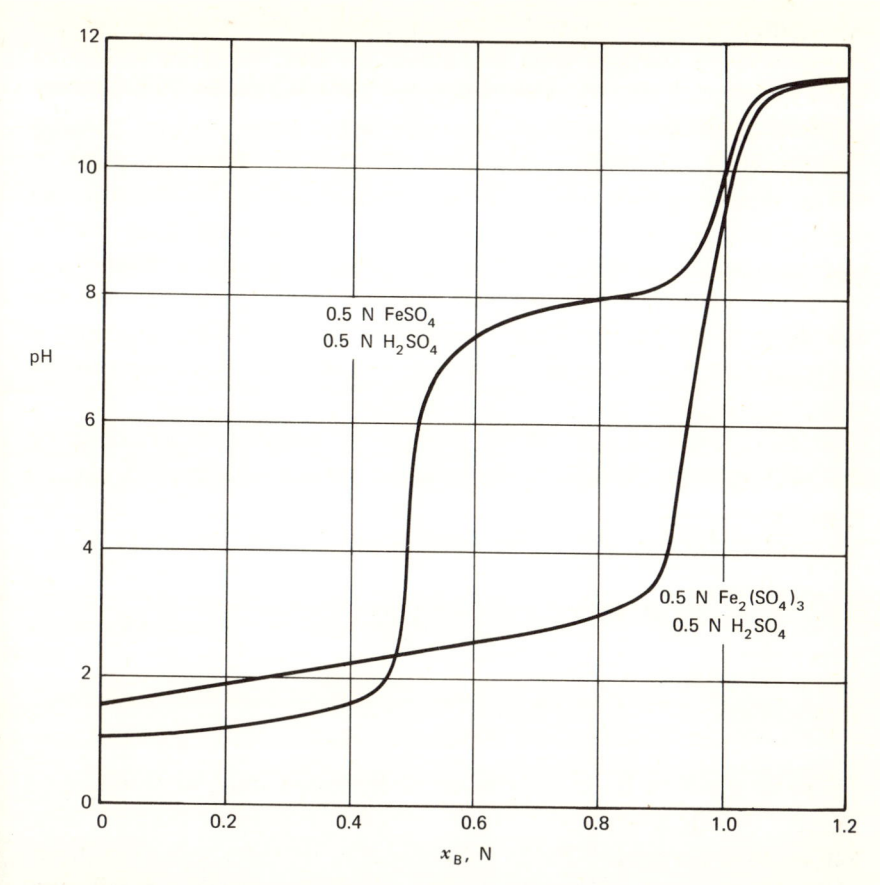

FIGURE 3.7. Titration curves for mixtures of ferrous and ferric sulfate with sulfuric acid.

of 2.5 and 4.7 for the two steps of the reaction:

$$Fe^{3+} + H_2O \rightleftharpoons FeOH^{2+} + H^+$$

$$FeOH^{2+} + H_2O \rightleftharpoons Fe(OH)_2^+ + H^+$$

The lower curve of Fig. 3.7 also bears this out. Again, precipitation occurs when the final hydroxyl group is added.

Mixtures of these metal ions from metal-finishing plants form titration curves which may be almost linear with a relatively low slope across most of the pH range. Neutralization of these wastes is surprisingly easy to control. The principal problem which may be encountered is variability in the titration curve. Absence of one or more metals due to processes which are not operating will affect the titration curve and possibly destabilize the control system. More is said about this in Chapter 9.

The sludge which settles to the bottom of a vessel following precipitation of these metal hydroxides acts as a huge reservoir for hydrogen and hydroxyl ions. Where it is not removed for weeks at a time, it can stabilize a system, for good or bad. If, for example, the discharge from a neutralization vessel to the sludge-settling tank is controlled in the pH range of 6 to 9, the effluent from the settling tank may hold within pH 7.5 to 8. However, if a major accident such as an uncontrolled caustic dump were to raise the pH entering the settling tank to 13 for an hour or so, the sludge will absorb hydroxyl ions extensively, releasing them gradually after control has been restored. The pH of the effluent may never exceed 11, but it may remain in the 9 to 10 range for a day or more after the accident, as the absorbed hydroxyl ions are released.

Applications of Hydrated Lime

There is much to be said in favor of hydrated lime, $Ca(OH)_2$, as a reagent for the adjustment of pH. It is less expensive than all other active basic reagents—Table 3.4 compares several reagents in terms of cost per ton of sulfuric acid neutralized. Ammonia is rarely used in waste-treatment applications since its salts are nutrients for plant growth, and its allowable concentration in effluents is strictly limited.

The solubility of hydrated lime is limited to 1.16 g/l at 25°C. If it were completely ionized, this solubility would give a pH of 12.64, whereas the actual pH of a saturated solution at 25°C is 12.53. It is a moderately strong base with pK_1 of 1.40 and pK_2 of 2.43. Its relatively low solubility makes it difficult to meter and control, but safe for workmen to handle. Suggestions for the manipulation of lime flow are given in Chapter 7.

Among its other advantages is the relative insolubility of many calcium salts. Calcium carbonate, sulfate, fluoride, and phosphate are all much less soluble than the same salts of sodium. Relatively large concentrations of these ions can thereby be removed simply

Table 3.4 Costs of Basic Reagents (1971 prices)

	Cost per Ton ($)	lb/lb H$_2$SO$_4$	Cost per Ton H$_2$SO$_4$ ($)
NaOH (solid, drums)	136.00	0.816	111.00
NaOH (50% solution, tank car)	71.00	1.632	116.00
Ca(OH)$_2$ (bags)	20.75	0.777	16.10
Na$_2$CO$_3$·H$_2$O (bags)	79.00	1.268	99.90
NH$_3$ (tank car)	67.50	0.347	23.42

by adding lime to a controlled pH. As an example, waste sulfuric acid from TNT production is neutralized with lime to precipitate the sulfate from solution.

The solubility of a salt can be expressed in a manner similar to ionization:

$$[Ca^{2+}][SO_4^{2-}] = K_{sp} \qquad (3.25)$$

where K_{sp} is the solubility product constant. It can be expressed as pK as was done with ionization constants: The pK for CaSO$_4$ is 4.2. If a stream containing only sulfuric acid is neutralized with lime, precipitation will not take place unless the ion product appearing in (3.25) exceeds $10^{-4.2}$. Since the concentration of both ions will be identical at neutrality,

$$[Ca^{2+}][SO_4^{2-}] = [SO_4^{2-}]^2$$

The minimum sulfate-ion concentration which will result in precipitation is

$$[SO_4^{2-}] = \sqrt{10^{-4.2}} = 10^{-2.1} = 0.8 \times 10^{-2} \, M$$

Expressed in normal units, this would be 1.6×10^{-2} N, corresponding to pH 1.87.

Fluorides are more easily precipitated:

$$[Ca^{2+}][F^-]^2 = 10^{-10.4}$$

Calcium carbonate is much less soluble than the sulfate:

$$[Ca^{2+}][CO_3^{2-}] = 10^{-8.1}$$

and phosphate is somewhat less soluble than the carbonate:

$$[Ca^{2+}]^3[PO_4^{3-}]^2 = 10^{-26}$$

$$[Ca^{2+}][HPO_4^{2-}] = 10^{-8.4}$$

[The relative solubility of various salts can be compared by dividing the pK by the sum of the coefficients of the ion-concentration terms. For $Ca_3(PO_4)_2$, this number is 26/5 or 5.2, and for $CaHPO_4$ it is 8.4/2 or 4.2. Thus the latter is an order of magnitude more soluble.]

The relative concentrations of the two phosphates present in a solution are a function of the pH. The pK values for phosphoric acid are 2.23, 7.21, and 12.32, so $Ca_3(PO_4)_2$ will only exist in significant concentration in very basic solutions. Butler[6] calculates $[PO_4^{3-}] = 1.68 \times 10^{-8}$ M and $[HPO_4^{2-}] = 3.04 \times 10^{-4}$ M at pH 8, so the more soluble HPO_4 is nonetheless the species more likely to precipitate. The total phosphate (PO_4) for the conditions above is 29 mg/l—far above federal standards.

As more lime is added, $[Ca^{2-}]$ and pH both increase, reducing the PO_4 level to 9 mg/l at pH 9 and 4 mg/l at pH 10 as indicated by tests on domestic sewage.[7] However, after adding lime to this level, neutralization by an acid is required to lower the pH again before discharging to a stream. Bear in mind, however, that pH is not a measure of calcium-ion concentration—acids or other bases in the stream profoundly affect the amount of lime required to reach a given pH. So pH control alone by lime addition cannot ensure an acceptable phosphate or fluoride level in the effluent unless the waste is known to contain only a very limited amount of acids and bases.

Obviously calcium chloride could be added in place of lime to precipitate these salts without resulting in an excessively high pH—in fact without affecting the pH at all. But in this instance, pH control cannot be used to establish the calcium level. The preferred method, whether lime or chloride is added, is calcium-ion control, which is investigated in Chapter 4. Control over pH would still be necessary to establish the PO$_4$–HPO$_4$ equilibrium at a desirable point.

NONAQUEOUS MEDIA

Reaction between acids and bases are often conducted in nonaqueous media as a part of some manufacturing process. Solvents which have been used include alcohols, ethers, hydrocarbons, ketones, and organic acids, as well as inorganic liquids such as anhydrous ammonia. Conventional pH electrodes may be used in the solvent if it is not completely anhydrous, and miscible with water. Methanol and acetone fit this category as long as a trace of water is present. If the solvent is anhydrous, however, it will dehydrate the glass membrane of the measuring electrode, raising its impedance to such a high level that a meaningful signal is unattainable.

Polar Solvents

Many solvents dissociate in a manner similar to water, producing one ion deficient in hydrogen and another with an excess. Anhydrous ammonia,[8] for example, dissociates as

$$NH_3 + NH_3 \rightleftarrows NH_4^+ + NH_2^-$$

Salts dissolved in such a solvent will act as strong acids or bases if they contain either of the ions formed by the solvent. Ammonium chloride in anhydrous ammonia will act as a strong acid like HCl in water; sodium amide acts as a strong base in ammonia.

Alcohols such as methanol ionize as follows:

$$CH_3OH + CH_3OH \rightleftarrows CH_3OH_2^+ + CH_3O^-$$

A glass electrode immersed in methanol containing neither acid nor base will indicate approximately pH 5. This is not necessarily the hydrogen-ion concentration or that of the positive ion given above, either. However, it is useful for control purposes, in that bases will raise the indicated pH and acids will lower it. A typical strong base in methanol is lithium methylate, $LiOCH_3$, and a typical weak base would be lithium boromethylate, $LiB(OCH_3)_4$. Trimethyl borate, $B(OCH_3)_3$, is a weak acid equivalent to boric acid in water. A 0.1 N mixture of $LiB(OCH_3)_4$ in methanol would indicate a pH slightly above 7; neutralization to indicated pH 5 with anhydrous HCl produces LiCl and Li_3BO_3. Further acidification to pH 1 will produce $B(OCH_3)_3$ along a polyprotic weak-acid titration curve.

Nonpolar Solvents

Ketones, ethers, and hydrocarbons do not dissociate freely into ions with proton excess or deficiency, yet acid-base reactions may be conducted in them. Boron trichloride, BCl_3, is a strong acid in solution with diethyl ether, although no hydrogen ion is dissociated from it. It reacts with the weak base lithium borohydride to form diborane gas:

$$BCl_3 + 3LiBH_4 \rightarrow 3LiCl\downarrow + 2B_2H_6\uparrow$$

That each reactant ionizes in solution is indicated by electrolytic conductivity proportional to their concentrations. The relatively insoluble products of the reaction do not noticeably conduct.

The endpoint of the reaction discussed above has been controlled at minimum conductivity, but not satisfactorily so. The difficulty comes in determining which reagent is in excess if the conductivity is high. A preferred method is to withdraw a sample, hydrolyze it, and measure the resulting pH.

Acid-base reactions may also be performed on organic chemicals in hydrocarbon solution. The reagents most commonly used for neutralization in these systems are anhydrous hydrogen chloride or anhydrous ammonia. Here again, a sample must be withdrawn and hydrolyzed with water before a pH measurement can be made. A scheme for separating the aqueous and organic layers is given in Fig. 7.8.

SUMMARY

A pH measurement can tell much about the state of a solution, whether the constituents are acids, bases, salts, or only partially soluble solids and gases. But obviously much more must be known and related to the pH measurement before the true state of the solution can be determined. In this chapter, some of the basic relationships have been derived as a starting point for those who wish to go further. And a variety of applications have been cited as examples of actual experiences which may be encountered in the field. Hopefully this combination will guide the engineer in this interpretation of some of his pH measurements and titration curves.

REFERENCES

1. D. M. Considine, *Process Instruments and Controls Handbook*, McGraw-Hill Book Company, New York, 1957, p. 12-6.
2. F. Daniels, *Outlines of Physical Chemistry*, John Wiley & Sons, Inc., New York, 1948, p. 460.
3. J. N. Butler, *Ionic Equilibrium: A Mathematical Approach*, Addison-Wesley, Reading, Massachusetts, 1964, p. 129.
4. W. A. Parsons, *Chemical Treatment of Sewage and Industrial Wastes*, National Lime Association, Washington, D.C., 1965, p. 25.
5. W. A. Parsons, *Ibid*, p. 56.
6. J. N. Butler, *op. cit.*, p. 507.
7. O. E. Albertson, and R. J. Sherwood, *Phosphate Extraction Process*, Dorr-Oliver, Inc., Stamford, Connecticut, 1968, p. 9.
8. J. N. Butler, *op. cit.*, p. 135.

4

OTHER ION-SELECTIVE MEASUREMENTS

The advent of ion-selective electrodes was something of an unusual development in engineering circles. Here were solutions looking for problems. The hydrogen-ion electrode had been with us for years, and in the laboratory at least were silver-ion electrodes. But all at once an entire array of sensitive electrodes were developed which were, and still are, looking for people to use them. None is, of course, as universally applicable as the hydrogen-ion electrode, because all of the process industries use acids and bases in some degree. The other electrodes may find selected usage only in particular industries, however, as silver in the photographic industry, cyanide in the plating industry, and so on. Rather than examining possible applications for each ion species or for each industry, it would seem more to the point to describe general types of applications, each illustrated with an example from industry.

MONITORING

When any new measuring device appears on the market, particularly one that has been developed without a definite application in mind, its first use will be simply as a monitor. This is certainly true of the ion-selective electrodes, especially since their arrival coincides with an awakening concern over stream pollution. Even as a monitor, their use extends far beyond waste streams. A few of the examples given below should spark the imagination of those engineers who have certain problems to solve and may not have thought of this means of solving them.

Effluent Discharge

The variety of toxic components flowing in plant effluents far outnumbers the available ion-selective electrodes—and unfortunately many of the worst offenders are not in an ionic state. However, many new electrodes are still under development, and perhaps soon a variety of heavy-metal sensors will be available. Objectionable ions which can be measured in waste streams at present are cyanide, cupric, lead, sulfide, and fluoride. The first two are finding application in the metal-plating industry, whose discharges are under close scrutiny because they use a variety of highly toxic chemicals. Fluorides are present in the wastes from aluminum plants while sulfides are principal contaminants in paper-mill effluents. Nitrate electrodes are already being used to monitor fertilizer runoff and nitrogen oxide emissions from combustion.[1]

Beyond the conventional uses of these electrodes in determining the source and intensity of ionic discharges, they can also be used for tracing. A salt of a relatively harmless ion like bromide could be used to determine dilution, leakage, or dynamic response of effluent systems or drains, by recording bromide-ion concentration at selected points throughout the system. Or a component whose presence is not detectable by any available means could be "tagged" by mixing with a small amount of a detectable ion, in much the same way as an odorizer is mixed with natural gas. Then the presence of the detectable ion would alert an operator to look for a malfunction or leak in his process.

Leak Detection

Processes using highly corrosive acids such as HCl or HF are quite prone to leakage. Care is taken in their design to keep process pressure above that of the cooling water so that leakage will be *into* the cooling-water system. (If water leaked into the process, vigorous reaction could occur with severe damage to equipment, while leakage in the other direction would ordinarily be less destructive.)

Leak detection with a pH electrode is not viable since cooling-water systems are generally pH-controlled. In addition, exposure to the atmosphere and addition of chemicals affect the pH. A fluoride-ion electrode is very sensitive, however, and could alarm the operator at the first detection of a leak in the HF system. Similarly a chloride electrode could be used to detect leaks into cooling-water or steam-condensate systems for processes using HCl or other corrosive chlorides.

Column Breakthrough

Ion-exchange columns present a natural application for selective electrodes. By far the most common use for ion exchange is in softening water by removal of the calcium and magnesium ions. When the column is approaching saturation, the level of these ions in the discharge will begin to rise. The divalent-ion or "water-hardness" electrode described in Chapter 1 may be used to automate switching and regeneration of the columns.[2]

Recovery of copper from plating or etching solutions is another case of breakthrough detection. In this process, however, copper may be plated onto scrap steel or machine-shop turnings. When the cupric-ion concentration in the effluent increases, it is a signal for the operator to replace the steel with a fresh lot. The accuracy of the measurement in this application is not absolute, however, due to a ferric-ion interference.

Ion exchange or precipitation with sulfide can also be used to remove copper from solution. Here again, a cupric-ion electrode can monitor the effluent and alarm on a failure or breakthrough. As copper continues to increase in cost, recovery operations like this will become commonplace.

Sodium-ion electrodes are now being used to detect the presence of this impurity in boiler feedwater. Sensitivity and selectivity are both important in this application, with measurements being made in the range of 0.1 to 1.0 ppb.[3] Iron is another contaminant whose presence is closely watched in feedwater, but which must be presently analyzed by other means.

Total Sulfide Calculations

Sulfide electrodes are being used to monitor the performance of oxidation units for black liquor in paper mills.[4] Usually a failure is indicated by an increase in sulfide-ion concentration by more than an order of magnitude. Where precise measurements must be made, however, due consideration must be given to the fact that the electrode sees only the sulfide ion. The weak acid, hydrogen sulfide, is easily formed, so sulfide can also exist as the HS^- ion and as H_2S in solution, depending on the pH. Total sulfide is the sum of the three:

$$x_S = [H_2S] + [HS^-] + [S^{2-}]$$

An expression for total acid (x_C) as H_2S was derived in Chapter 3. The total sulfide (as H_2S) is half that value:

$$x_S = [S^{2-}][1 + 10^{pK_2 - pH}(1 + 10^{pK_1 - pH})] \qquad (4.1)$$

If total sulfide is the measurement of importance, pH must either be controlled or compensated.

Since the sulfide-ion measurement is made on a logarithmic scale as pS^{2-}, (4.1) will be more useful if expressed in those terms. Here pS will indicate the negative $\log_{10}$ of total sulfide concentration:

$$pS = pS^{2-} - \log[1 + 10^{pK_2 - pH}(1 + 10^{pK_1 - pH})] \qquad (4.2)$$

A plot of the correction factor given in (4.2) appears in Fig. 4.1. Observe that the correction factor is 1 pS/pH between pH 7 and 13 and 2 pS/pH below pH 7. (The pK values for H_2S are 7.00 and 12.92.)

In situations where the pH cannot readily be controlled, compensation for its variation can be applied using the analog computing scheme shown in Fig. 4.2. The function generator develops the curve of Fig. 4.1 while the subtracting unit applies the correction. Note the common reference electrode.

Sulfur can also exist as the element itself, in which case it can neither be detected by the electrode nor be accounted for by pH. In sulfide solutions of moderate pH, simple exposure to the atmosphere is sufficient to cause oxidation of sulfides to elemental

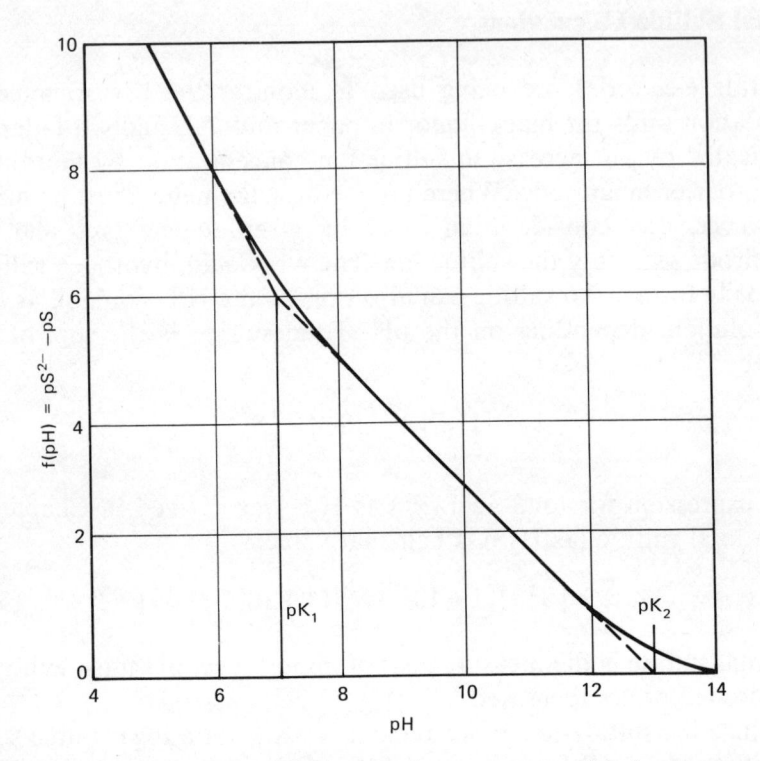

FIGURE 4.1. The function of pH required to convert a sulfide-ion measurement to total sulfide is here plotted against pH.

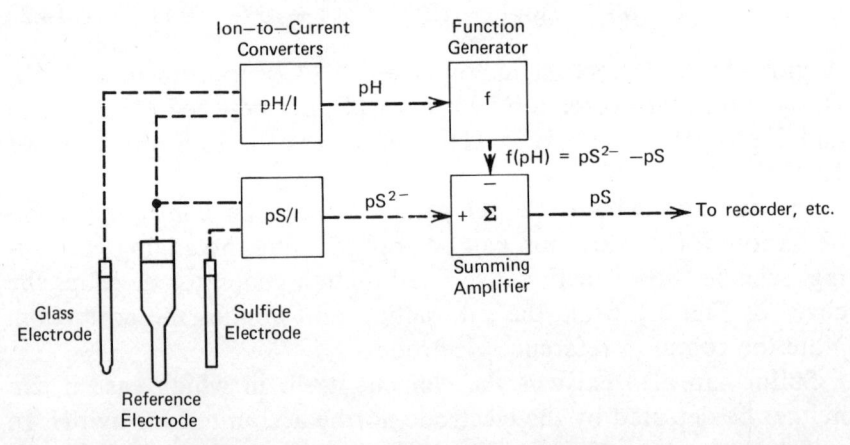

FIGURE 4.2. Total sulfide concentration can be calculated from pH and sulfide-ion activity.

84

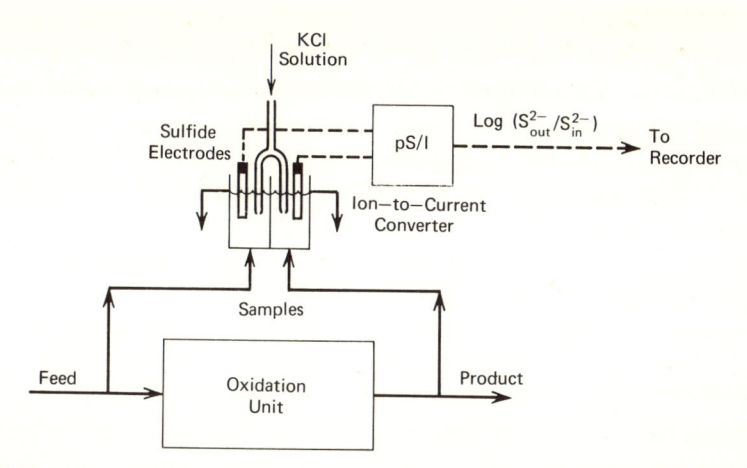

FIGURE 4.3. A differential ion measurement can indicate the efficiency of a processing unit.

sulfur. Another loss of total sulfide is the possible escape of H$_2$S at pH levels below 6.

If the feed and discharge from an oxidation unit or ion-exchange column are at essentially the same pH and temperature, a differential measurement can be made between sensitive electrodes in each stream. The differential, $pS^{2-}_{out} - pS^{2-}_{in}$, indicates the efficiency of the unit, since the subtraction of logarithms is a dividing operation:

$$pS^{2-}_{out} - pS^{2-}_{in} = \log \frac{[S^{2-}_{out}]}{[S^{2-}_{in}]} \qquad (4.3)$$

A system for making this measurement is shown in Fig. 4.3; no reference electrode is used, but a salt bridge is necessary to provide conduction between the two samples.

BLENDING

Blending represents the simplest form of continuous control in which selective-ion electrodes might be used. The flow of reagent is simply modulated to bring the ion concentration to the desired set point. The only obstacles in the way are the limited accuracy and the acute nonlinearity of the measurement.

As stated before, pIon measurements are not terribly accurate: ±1 mV at best, equivalent to ±3.8% of value. In blending situations, precise control over concentration is desirable, while sensitivity and rangeability are not as important. The minimum range available in a standard pIon amplifier is 50 mV. If it is desired to control a caustic solution at 10%, for example, the range of the instrument could be one decade—3 to 30% caustic, for example—encompassing more than is either necessary or desirable. The accuracy of ±1 mV could correspond to ±0.38% caustic; a more commonly attainable accuracy of ±6 mV would correspond to ±2.2% caustic. But at this concentration, there are many more sources of error. The liquid-junction potential is large, and the activity coefficient quite variable. In addition, a pH electrode has a high sodium interference in this range. The sodium-ion electrode would be more effective, but if other anions like chloride were present, this measurement would not describe the actual hydroxide concentration. But, all things considered, ion potential is a poor choice for this application—electrolytic conductivity, though nonspecific, is more accurate in this range.

Divalent electrodes are half as accurate as univalent electrodes, since they generate half the number of millivolts per decade, and errors in a measuring system are millivolt errors. Yet, in many cases, the ion-selective electrodes are still the best choice for a particular application.

Fluoridation

Automatic control of water fluoridation can be achieved with the fluoride-ion electrode. If water supplies were fluoride free, dosage could be controlled on a measured basis. But fluoride content in a raw water is significant and variable, frequently requiring fluoride-ion control.

The fluoride-ion electrode is filled with an internal solution containing 1 ppm of fluoride ion. This activity coincides with the desired operating level for fluoridation systems; operation at this point then eliminates the need for temperature compensation. The range of the instrument is usually selected as 0.1 to 10 ppm, a two-decade logarithmic scale.

Since the fluoride ion is negative, increasing concentration develops a negative voltage, with 0 mV at center scale. At this level of concentration and pH, accuracy is good.

Water-Hardness Control

It is not always desirable to reduce water hardness to an absolute minimum. In fact, very soft water tends to be quite corrosive to certain metals, particularly aluminum and zinc. The water softened by ion-exchange columns is not of a constant hardness—its ion concentration varies with the raw-water hardness, the velocity of flow, and the loading of absorbed ions on the bed. To deliver water of a uniform hardness to processing equipment, deionized water (containing some residual hardness) can be blended with raw water. A water-hardness electrode is used to measure the condition of the blend, with control exercised over the addition of raw water, as shown in Fig. 4.4.

Water hardness is generally determined analytically by a soap test, and is reported as parts per million equivalent calcium carbonate, although magnesium and, to a lessser extent, zinc and iron also cause hardness. The divalent-ion electrode contains a liquid ion exchanger responsive equally to Ca^{2+}, Mg^{2+}, and Ba^{2+}, 1.3 times as sensitive to Ni^{2+}, and 3.5 times as sensitive to Zn^{2+} and Fe^{2+}. But since the first two species predominate in most water sources, the

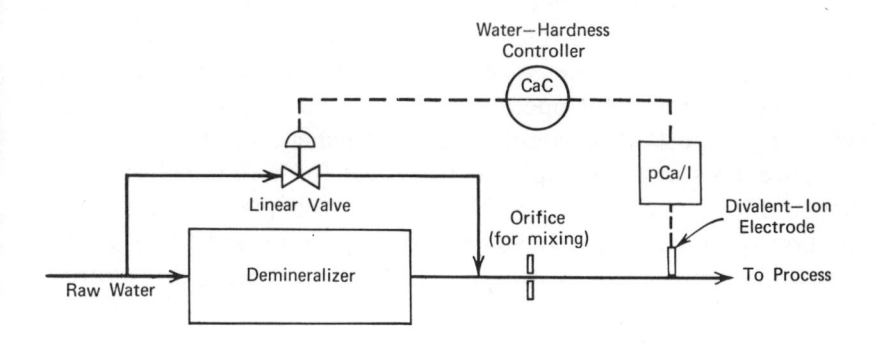

FIGURE 4.4. Blending of raw and demineralized water may be controlled using a divalent-ion electrode.

error introduced by those ions with higher sensitivities is not significant.

Since the principal constituents of water hardness respond exactly as Ca^{2+}, calibration of the electrode in terms of ppm $CaCO_3$ is possible, and in fact customary. A typical range would be 1 to 100 ppm in two logarithmic decades. (Single decade calibration—approximately 30 mV—is below the span limits for a standard instrument.)

Increased emphasis will be placed on water-hardness control in the future, as chemical treatment methods involving liming under close pH control come into practice. In these processes, calcium and magnesium are precipitated as carbonate and hydroxide, respectively, by raising the pH to 11 with lime. After clarification, readjustment of the pH to 8 by carbonation can yield a high-quality water of controlled hardness. Final hardness is established by the Mg^{2+} content of the raw water and the exact pH reached by liming.

Scrubbing Solutions

Caustic soda has long been used to scrub acid gases from vent systems in chemical plants. Precise control of caustic concentration was not important since an excess was desirable. Regeneration was seldom considered.

Now, however, caustic scrubbing solutions are being used to remove sulfur dioxide from the flue gases of industrial and central-station power plants. Since 1000 MW of electrical generation can produce as much as 600 tons of sulfuric acid per day of continuous operation, huge quantities of chemicals will be used for scrubbing. Economic and efficient operation of the scrubbing system becomes extremely important, particularly since regeneration is necessary.

Figure 4.5 shows a simplified flowsheet for a flue gas scrubbing system. Gas entering the scrubber comes in contact with a caustic solution of controlled pH. Sulfur dioxide is removed from the flue gas, while some water is evaporated and leaves with the scrubbed gas:

$$SO_2 + NaOH \rightarrow NaHSO_3 \qquad (4.4)$$

Spent solution is then transferred to the regenerator where it is

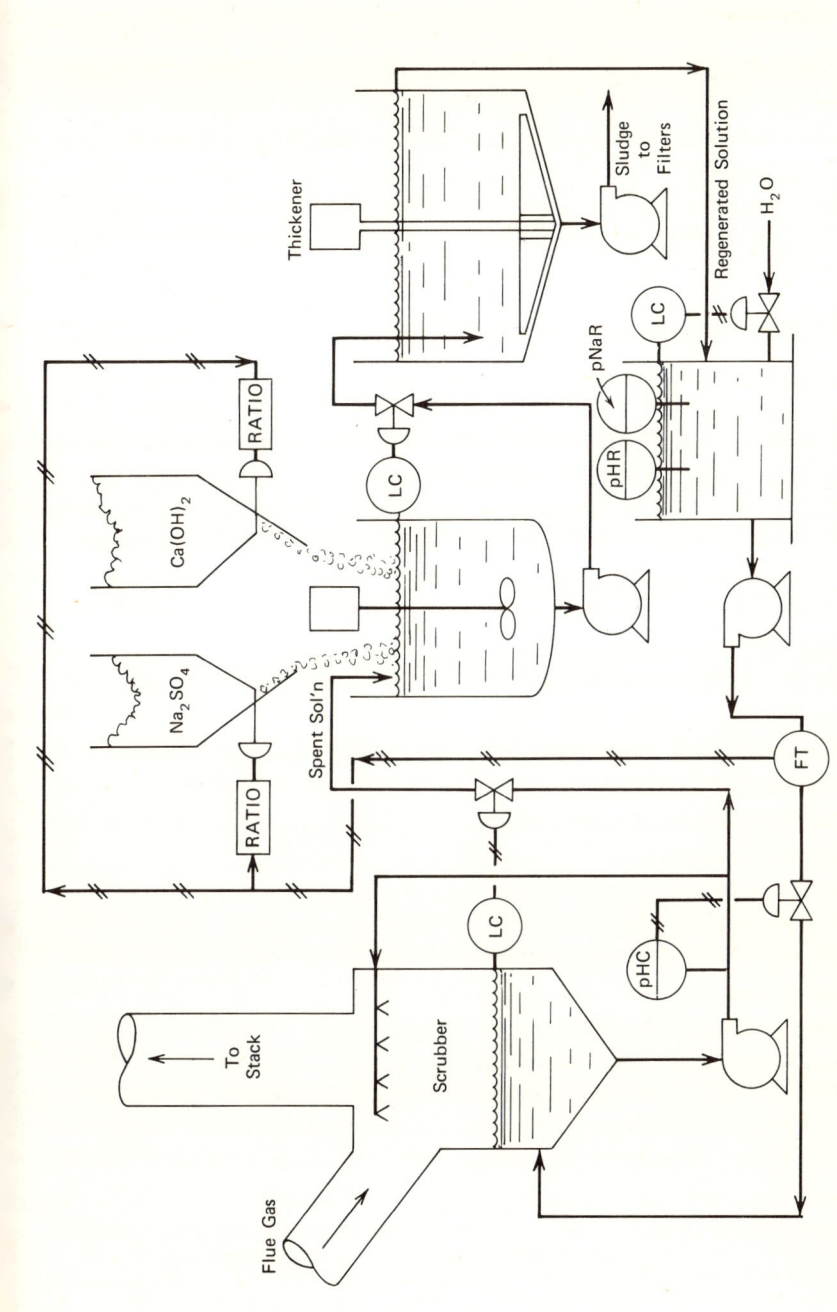

FIGURE 4.5. Lime and sodium sulfate must be added in proportion to the consumption of scrubbing solution by the flue gas.

mixed with lime, regenerating the caustic while precipitating calcium sulfite:

$$NaHSO_3 + Ca(OH)_2 \rightarrow NaOH + CaSO_3\downarrow \qquad (4.5)$$

The latter is removed as a sludge, while the supernatant caustic is recycled to the scrubber.

Oddly enough, a saturated solution of lime has a maximum pH of only 12.53. But because calcium sulfite is less soluble (0.0043 g/100 g H_2O compared to lime at 0.116 g/100 g), calcium ions are taken from the lime by the sulfite, leaving the hydroxyl ions free. Calcium-ion concentration in the regenerated solution is only about 10^{-4} M, owing to a high sulfite-ion concentration and the low solubility of $CaSO_3$. Sodium-ion concentration is about 1 M while hydroxyl-ion concentration is held near 0.5 N. Scrubbing solution pH is controlled by lime dosage, but it is too high to measure accurately, as stated earlier. Nonetheless it can be regulated indirectly in conjunction with scrubber pH.

The pH in the scrubber must be precisely controlled at 6.0—if it is too low, SO_2 will pass unabsorbed; if too high, CO_2 will be absorbed in addition. (The pK values of SO_2 are 1.8 and 5.3, compared to 6.35 and 10.25 for CO_2.) Needless to say, there is much more CO_2 than SO_2 in the flue gas. Absorption of CO_2 will consume equivalent quantities of lime, with the ensuing precipitation of $CaCO_3$. Therefore increasing the pH in the scrubber could result in lime consumption all out of proportion to the SO_2 in the flue gas.

The scrubber pH controller adds scrubbing solution as necessary to match the rate of SO_2 absorption. This raises the level in the scrubber, causing the level controller to withdraw spent solution at an equivalent rate. Lime is added in ratio to the flow of scrubbing solution—if no scrubbing solution is needed to satisfy the scrubber pH, no lime is needed. The ratio between lime flow and scrubbing solution flow ultimately determines the OH^- content of the solution.

This system is completely self-regulating. If the operator should increase the lime-to-scrubbing solution ratio, the OH^- level of the solution would increase, but not without limit. A higher OH^- level in the scrubbing solution will mean that a lower flow will satisfy a given rate of SO_2 absorption. The lower flow will therefore lower the lime addition rate until it again balances the SO_2 absorption rate at the now higher OH^- level.

Sodium ions must also be added, but only in an amount sufficient to make up for losses, principally in the solution trapped with the wet sludge. As with the lime flow, sodium sulfate is added in ratio to the usage of scrubbing solution. The value of the ratio setting establishes the level of sodium ions in the solution. Sodium-ion concentration is then also regulated, since a higher concentration will increase the rate of sodium loss proportionately.

Sodium-ion and pH measurements are made on the scrubbing solution. Their accuracy is relatively poor at the concentrations given, but trends will be noticeable and abnormal operation revealed. Calibration against laboratory analysis is essential.

Example 4.1

Establish the ion concentrations for the regeneration of Na_2SO_3 by liming:

$$Na^+ + SO_3^{2-} + Ca^{2+} + OH^- \rightleftharpoons CaSO_3\downarrow + Ca(OH)_2\downarrow$$

The charge balance is

$$[Na^+] + 2[Ca^{2+}] = 2[SO_3^{2-}] + [OH^-]$$

Let $[Na^+] = 1.0\ M$. Then,

$$[OH^-] + 2[SO_3^{2-}] - 2[Ca^{2+}] = 1$$

Solubility product constants are

$$[Ca^{2+}][SO_3^{2-}] = K_s = 1.65 \times 10^{-5}$$

$$[Ca^{2+}][OH^-]^2 = K_0 = 3.2 \times 10^{-5}$$

[The last statement assumes complete dissociation of $Ca(OH)_2$, for purposes of simplification.] Substituting K_s and K_0 in the charge balance yields

$$[OH^-] + 2\frac{K_s}{K_0}[OH^-]^2 - 2\frac{K_0}{[OH^-]} = 1$$

Trial-and-error solution of the last equation by substituting selected values of [OH$^-$] yields the following results:

$$[Na^+] = 1.0\ M \qquad [SO_3^{2-}] = 0.1\ M$$

$$[OH^-] = 0.6\ N \qquad [Ca^{2+}] = 8.4 \times 10^{-5}\ M$$

The OH$^-$ level thus found is the highest achievable at the selected Na$^+$ activity, since it assumes equilibrium with solid lime. To avoid lime losses, either Na$^+$ must be raised or OH$^-$ reduced by adjustment of the ratio controls. Concentrations measured in the actual plant will differ from those listed above due to the presence of sulfate and carbonate ions as well as undissociated and partially dissociated lime.

ION REMOVAL

Chemical reactions tend to proceed in the direction of the lowest energy state, like stones rolling downhill. Acid-base reactions, for example, combine two highly active species to produce that very inert, barely dissociated substance called water. There are other types of ionic, liquid-phase reactions, and they too go in the direction of comparatively inert products. Ions may react to form a soluble but only slightly ionized product, or slightly soluble products—precipitates and gases.

In acid-base reactions, hydrogen-ion concentration was the key variable to indicate the state of the system. Where precipitates or gases are generated, other ion measurements may be used for control, unless the products of the reaction contain H or OH groups, as, for example, H$_2$S gas or Fe(OH)$_3$ solid. This introduces us to a new class of reactions which are different from acid-base reactions, yet follow many of the same principles.

Solubility Titration Curves

The relationship between H$^+$ and OH$^-$ ions in an aqueous solution is fixed by the dissociation constant of water:

$$[H^+][OH^-] = K_w \qquad (4.6)$$

Restating in p terms:

$$pH + pOH = pK_w = 14 \quad \text{at} \quad 25°C \quad (4.7)$$

By the same token, certain ions combine to form products of limited solubility. The equilibrium constant relating their activities is the solubility product constant. The reaction between silver and bromide ions is typical:

$$[Ag^+][Br^-] = K_B \cong 10^{-12} \quad \text{at} \quad 25°C \quad (4.8)$$

Again, restating in p terms:

$$pAg + pBr = pK_B \cong 12 \quad (4.9)$$

Recognize the similarity between forming a precipitate and neutralizing an acid or base. Silver nitrate and sodium bromide solutions may be mixed as was done with HCl and NaOH in the previous chapter:

$$AgNO_3 + NaBr \rightarrow Ag^+ + NO_3^- + Na^+ + Br^- + AgBr\downarrow \quad (4.10)$$

All of the nitrate ions were charged with the silver nitrate, and all of the sodium ions entered with the bromide. Let their initial concentrations be $x_A = NO_3^-$ and $x_B = Na^+$. A charge balance is

$$[Ag^+] + [Na^+] = [NO_3^-] + [Br^-]$$

Substituting x_A and x_B:

$$x_A - x_B = [Ag^+] - [Br^-]$$

Substituting for $[Br^-]$ from (4.8) yields

$$x_A - x_B = [Ag^+] - \frac{10^{-12}}{[Ag^+]} \quad (4.11)$$

Equation (4.11) may be restated in p units:

$$x_A - x_B = 10^{-pAg} - 10^{pAg-12} \quad (4.12)$$

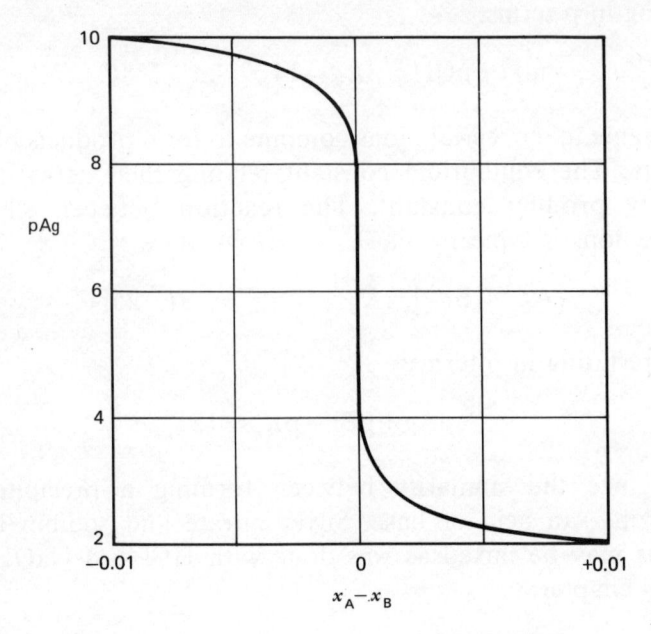

FIGURE 4.6. The titration curve of silver against bromide ion mimics an acid-base neutralization.

Note that the form of (4.12) is identical to that of (3.5)—the only differences are the use of pAg instead of pH and 12 instead of 14.

A titration of silver against bromide ion is remarkably similar to strong acid against strong base. The nonlinearity, the sensitivity, the extreme accuracy required at the equivalence point are virtually the same, as Fig. 4.6 demonstrates. Equivalence takes place where [Ag$^+$] and [Br$^-$] are equal, that is, at pK_B/2 or pAg 6. The same rules apply to other salts of limited solubility.

If sodium chloride were added to the solution above, two precipitates would be formed:

$$AgNO_3 + NaBr + NaCl \rightarrow Ag^+ + NO_3^-$$
$$+ Na^+ + Br^- + Cl^- + AgBr\downarrow + AgCl\downarrow \quad (4.13)$$

Since AgCl is the more soluble of the two, it will not precipitate until a certain Ag$^+$ activity is reached, below which point the titration curve described by (4.12) still applies. However, when its solubility limit is reached, AgCl precipitation controls the titration. Let x_C represent the concentration of NaCl introduced, and K_C the solubility product of AgCl. Where $[Ag^+] < x_C/K_C$, (4.12) applies. For greater values of $[Ag^+]$, a new charge balance must be written:

$$[Ag^+] + [Na^+] = [NO_3^-] + [Br^-] + [Cl^-]$$

Substituting x_A for $[NO_3^-]$ and $x_B + x_C$ for $[Na^+]$ yields the following:

$$x_A - x_B - x_C = [Ag^+] - \frac{K_B}{[Ag^+]} - \frac{K_C}{[Ag^+]} \qquad (4.14)$$

Restating in p terms:

$$x_A - x_B - x_C = 10^{-pAg} - 10^{pAg-pK_B} - 10^{pAg-pK_C} \qquad (4.15)$$

Again, where precipitation of AgCl does not take place, the last term in (4.14) and (4.15) is x_C. In other words, x_C is the maximum value of that term.

Example 4.2

Derive the titration curve for the double precipitation described above, between the limits of pAg 2 to 10 (see the accompanying table), using $pK_B = 12$, $pK_C = 10$, and $x_C = 0.01$. (The maximum value of 10^{pAg-10} is then 0.01.)

Table 4.1

pAg	10^{-Ag}	10^{pAg-12}	10^{pAg-10}	$x_A - x_B - x_C$
10	10^{-10}	10^{-2}	10^{-2}	-0.02
9	10^{-9}	10^{-3}	10^{-2}	-0.011
8	10^{-8}	10^{-4}	10^{-2}	-0.0101
7	10^{-7}	10^{-5}	10^{-3}	-0.001
6	10^{-6}	10^{-6}	10^{-4}	-0.0001
5	10^{-5}	10^{-7}	10^{-5}	-10^{-7}
4	10^{-4}	10^{-8}	10^{-6}	$+0.0001$
3	10^{-3}	10^{-9}	10^{-7}	$+0.001$
2	10^{-2}	10^{-10}	10^{-8}	$+0.01$

96 Other Ions

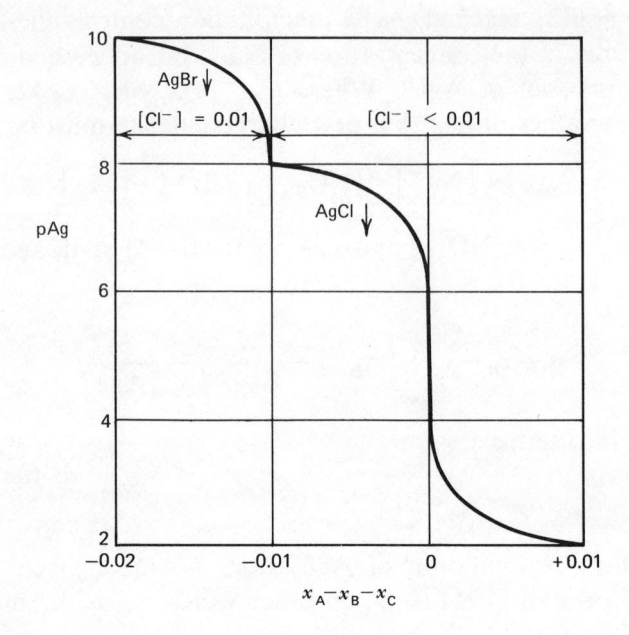

FIGURE 4.7. This titration curve describes the sequential precipitation of AgBr and AgCl from a solution with an initial chloride-ion activity of 0.01 M.

Figure 4.7 is a plot of the titration curve derived in Example 4.2. It is essentially a superposition of the bromide and chloride titration curves, with both inflections clearly distinguishable. Nonetheless, this titration curve does not resemble those of weak acids or bases, primarily because of the difference between solubility and dissociation equilibria. The solubility product constant includes no term representing the concentration of the solid, while the dissociation constant does include the concentration of the un-ionized acid or base [see (3.8)].

Companion-Ion Control

Just as a pH electrode infers hydroxyl-ion activity by means of the dissociation constant of water, an unmeasured ion activity may be inferred through the solubility product constant which relates it to the ion being measured. Thus the silver sulfide electrode is sensitive to silver ions directly and to sulfide ions through the K_{sp} of Ag_2S (10^{-51}).

The silver wire with silver chloride coating was primarily an Ag^+ electrode, but would indicate Cl^- activity as well, in a solution saturated with AgCl. The Ag_2S–AgX pressed pellets do not have to meet this constraint. The silver sulfide acts simply as a matrix for the silver halide crystals, such that the electrode is sensitive to the halide ion. If, however, the electrode is used in a solution containing an ion of a less soluble silver salt, the halide in the electrode will be replaced by that salt. An Ag_2S–AgCl electrode in a solution of Br^- ions, for example, will eventually be converted to an Ag_2S–AgBr electrode with a commensurate shift in potential.

The Ag_2S–AgI electrode may be used for CN^- measurement, but the lower solubility of AgCN eventually consumes the electrode. Again, the electrode is primarily an Ag^+ measuring device, indicating CN^- by means of its K_{sp}.

Silver sulfide can also act as a matrix for other metal sulfides of low solubility such as CuS, CdS, and PbS.[5] Again, these are primarily Ag^+ electrodes, but can indicate the other metal ion through the K_{sp} of both sulfides. For example,

$$[Ag^+]^2[S^{2-}] = K_{Ag}$$

$$[Cu^{2+}][S^{2-}] = K_{Cu}$$

Using S^{2-} as common for both salts, $[Ag^+]$ can be evaluated in terms of Cu^{2+}:

$$[Ag^+] = \frac{K_{Ag}}{K_{Cu}}[Cu^{2+}]^{1/2} \qquad (4.16)$$

Obviously silver ion acts as an interference with these mixed-sulfide electrodes.

The best selectivity for companion-ion measurement is achieved with the least soluble substances. An excellent example of this relationship is the use of a sulfide-ion measurement to control mercury concentration in the waste from chlorine alkali plants. Mercuric sulfide is one of the least soluble substances known, with a K_{sp} estimated at 10^{-50}. Control over sulfide-ion level therefore can precipitate mercury quite effectively—however, certain constraints apply.

98 Other Ions

As with many insoluble materials, some conditions can be found where they are quite soluble. Mercuric sulfide is one of these—it forms soluble polysulfides in highly alkaline solutions. This is unfortunate, since the sulfide ion is most completely dissociated under these conditions.

At the other end of the pH spectrum, in acid solutions, much of the sulfide in a solution is tied up as HS^- or H_2S. Since only S^{2-} will precipitate mercury, a reaction carried out at a low pH will consume excessive sulfide reagent, while generating hydrogen sulfide, another pollutant. Equation (4.2) and Fig. 4.1 define the difference between total sulfide in the system and the activity of free sulfide.

Mercury precipitation is customarily carried out by adding NaHS reagent to the waste at a controlled pH near 7.0. At this pH, Fig. 4.1 shows that the NaHS concentration must be 2×10^6 times as great as the S^{2-} activity required for precipitation. Controls for this reaction are illustrated in Fig. 4.8.

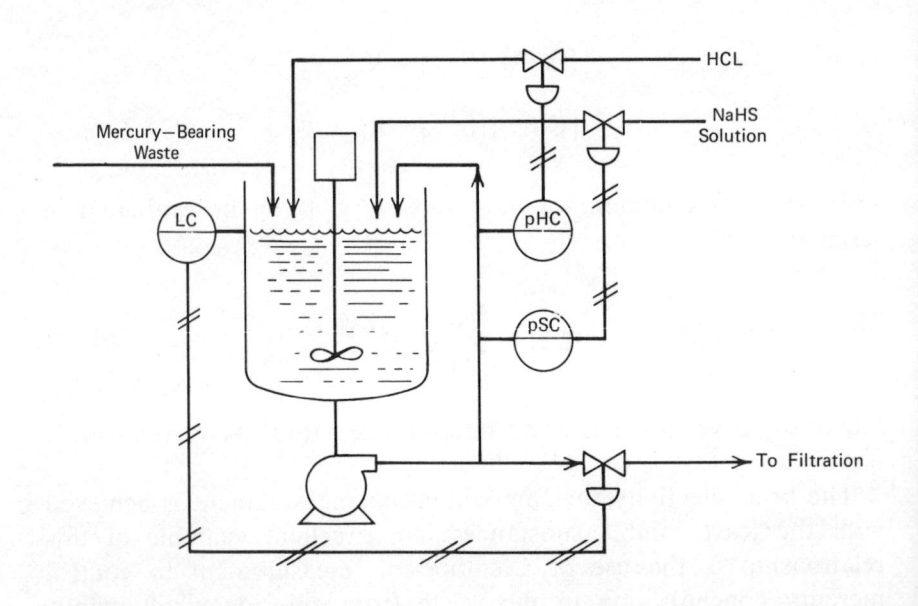

FIGURE 4.8. Mercury precipitation requires close control over both pH and sulfide-ion activity.

Complexation

When ammonia is added to a solution of cupric ions, the hydroxide is first formed:

$$Cu^{2+} + 2NH_3 + 2H_2O \rightarrow Cu(OH)_2\downarrow + 2NH_4^+$$

However, further addition of ammonia redissolves the precipitate to form a deep-blue solution characteristic of the copper-ammonia complex:

$$Cu(OH)_2\downarrow + 4NH_3 \rightarrow [Cu(NH_3)_4]^{2+} + 2OH^-$$

The complex copper-ammonia ion exhibits none of the properties of either the cupric ion or ammonia. Consequently a cupric-ion electrode would not measure its presence, although a pH electrode would indicate the OH$^-$ ion activity.

The unusual solubility of AgCl in concentrated KCl solutions was discussed in Chapter 1—this is another example of complexation. Silver-ion activity was fixed by the chloride-ion activity through the K_{sp}, however, indicating that the complex contributed no silver ions.

Another example is the dissolution of iodine in potassium iodide solution:

$$I^- + I_2 \rightarrow I_3^-$$

The triiodide ion is a complex which does not behave like the iodide ion.

Complex ions are generally very stable, only weakly dissociating into their components, as, for example, the argenticyanide complex:

$$[Ag(CN)]^- \rightleftharpoons Ag^+ + 2CN^-$$

The ionization constant for this equilibrium is

$$K = \frac{[Ag^+][CN^-]^2}{\left[Ag(CN_2)^-\right]} = 10^{-22} \qquad (4.17)$$

The ferrocyanide ion $[Fe(CN)_6]^{-4}$ and the ferricyanide ion $[Fe(CN)_6]^{-3}$ are very familiar complex ions of iron. They too barely dissociate into iron and cyanide ions, so that they do not react like iron salts, and are not toxic as cyanides are.

REFERENCES

1. J. N. Driscoll, et al., Determination of Oxides of Nitrogen in Combustion Effluents with a Nitrate Ion-Selective Electrode, 64th Annual Meeting of the Air Pollution Control Association, June 27, 1971.
2. R. T. Oliver and R. F. Mannion, Ion-Exchange Electrodes in Process Control: Water-Hardness Measurement of Ion-Exchange Treated Water, Presented at Analysis Instrumentation Symposium of the Instrument Society of America, May 25, 1970.
3. W. A. Lower and E. L. Eckfeldt, Sodium-Ion Monitoring, *Industrial Water Engineering*, March 1969.
4. J. L. Swartz and T. S. Light, Analysis of Alkaline Pulping Liquor with Sulfide Ion-Selective Electrode, *Tappi*, January 1970.
5. R. A. Durst, *Ion-Selective Electrodes*, National Bureau of Standards Special Publication 314, 1969, p. 79.

5

REDUCTION-OXIDATION MEASUREMENTS

The reactions wherein electrons are transferred between ions, radicals, and elements are known as oxidation and reduction reactions. In each of these, one component is reduced by gaining one or more electrons while another is oxidized by losing them. Metallic iron, for example, is oxidized by air to the Fe^{3+} state with the loss of three electrons, while oxygen is simultaneously reduced to O^{2-} with the gain of two electrons. Iron may also be oxidized from the Fe^{2+} to the Fe^{3+} state by oxygen, chlorine, permanganate, or similar oxidizing agent. It is basically an electrolytic reaction, with the exchange of electrons becoming a current flow.

The state of the reaction may be measured by the potential developed between an inert and a reference electrode. The potential thus measured is known variously as an oxidation-reduction potential (ORP) or reduction-oxidation potential (redox) measurement.

REDUCTION REACTIONS

The European convention, which is fast becoming an international convention, uses reduction potentials, whereas American usage has primarily been restricted to oxidation potentials. The only difference between the two is the sign given to the potential. Reduction potentials are used here.

Reduction Potentials

As described in Chapter 1, the standard hydrogen electrode has arbitrarily been assigned a potential of zero. At the surface of this (platinum) electrode, hydrogen gas is in equilibrium with hydrogen ions at unit activity in the surrounding (HCl) solution. The electrode develops a voltage relative to the state of the reaction:

$$E = E_0 + \frac{2.3RT}{F} \log \frac{[H^+]}{[H^\circ]} \quad (5.1)$$

Since E is defined as zero, gases and solids are arbitrarily assigned a unit activity, and $[H^+]$ is fixed at 1.0 by the makeup of the solution; E_0 is also zero.

Equation (5.1) is the Nernst equation for the reduction of hydrogen ions. The oxidized state of the ion (H^+) is in the numerator, the sign of the logarithmic term is positive, and E_0 is a *reduction* potential. An abbreviated list of standard reduction potentials is given in Table 5.1 to illustrate the relative driving forces for reactions.

Note the position of the metal ions in the table with respect to hydrogen—it is the familiar electromotive series of elements. The other species appear in an order relative to their oxidizing capability — ozone, hydrogen peroxide, and permanganate being strong oxidizing agents, while sulfite and nitrite ions are effective reducing agents. Species differing markedly with one another in reduction potential will react, and greater differences generally mean more vigorous reactions. Chromates, for example, will oxidize (or be reduced by) ferrous ions, metallic iron, or sulfite ions. And as will be seen, hypochlorite (ClO^-), hydrogen peroxide, and ozone react vigorously with cyanide.

As one species is reduced, another is oxidized. Although some of the equations in Table 5.1 seem to be written backwards, in that the reaction normally proceeds in the other direction, they are all reduction equations. Written in the reverse direction they would be oxidation equations, and their potentials will be reversed in sign.

Titration Curves

Nernst equations for inert electrodes express the relationship between measured potentials in terms of the ratio of the oxidized to

Table 5.1 Some Common Reduction Potentials[a]

Reduction Equation	E_o (mV)
$O_3 + 2H^+ + 2e \rightarrow O_2 + H_2O$	+2070
$H_2O_2 + 2H^+ + 2e \rightarrow 2H_2O$	+1776
$MnO_4^- + 8H^+ + 5e \rightarrow Mn^{2+} + 4H_2O$	+1490
$Cr_2O_7^{2-} + 14H^+ + 6e \rightarrow 2Cr^{3+} + 7H_2O$	+1330
$O_2 + 4H^+ + 4e \rightarrow 2H_2O$	+1229
$ClO^- + H_2O + 2e \rightarrow Cl^- + 2OH^-$	+890
$Ag^+ + e \rightarrow Ag$	+800
$Fe^{3+} + e \rightarrow Fe^{2+}$	+771
$Fe^{3+} + e \rightarrow Fe^{2+}$ (1 M H_2SO_4)	+685
$Fe(CN)_6^{3-} + e \rightarrow Fe(CN)_6^{2-}$	+440
$O_2 + 2H_2O + 4e \rightarrow 4OH^-$	+401
$Cu + 2e \rightarrow Cu^{2+}$	+345
$SO_4^{2-} + 2H^+ + 2e \rightarrow SO_3^{2-} + H_2O$	+170
$Sb_2O_3 + 6H^+ + 6e \rightarrow 2Sb + 3H_2O$	+145
$NO_3^- + H_2O + 2e \rightarrow NO_2^- + 2OH^-$	+10
$2H^+ + 2e \rightarrow H_2$	0
$Cd^{2+} + 2e \rightarrow Cd$	−402
$Fe^{2+} + 2e \rightarrow Fe$	−440
$S + 2e \rightarrow S^{2-}$	−508
$Zn^{2+} + 2e \rightarrow Zn$	−763
$CNO^- + H_2O + 2e \rightarrow CN^- + 2OH^-$	−970
$Al^{3+} + 3e \rightarrow Al$	−1706
$Na^+ + e \rightarrow Na$	−2711

[a] Most of the potentials were taken from Reference 1.

the reduced species. In a solution wherein no reaction is taking place, as in the redox standard solution given in Table 2.1, there is but one Nernst equation describing the system. Ordinarily, however, potential measurements are not used simply to monitor solutions, but to control the state of a reaction. And each reaction is really a pair of reactions—an oxidation and a reduction—each developing a potential represented by its own Nernst equation.

Consider an oxidizing agent which reacts as follows:

$$Ox^+ + 2e \rightarrow Ox^-$$

Reduction-Oxidation

and a reducing agent described similarly:

$$Re^+ + 2e \rightarrow Re^-$$

The potential developed by the equilibrium between Ox^+ and Ox^- is

$$E = E_{Ox} + \frac{2.3RT}{2F} \log \frac{[Ox^+]}{[Ox^-]} \qquad (5.2)$$

and that developed by Re^+ and Re^- is

$$E = E_{Re} + \frac{2.3RT}{2F} \log \frac{[Re^+]}{[Re^-]} \qquad (5.3)$$

Let the solution initially contain the reductant alone at an activity $[Re^-]$ identified as total reductant $[Re]$, with no $[Re^+]$ present. Then oxidant will be gradually added to a total activity level of $[Ox]$, with the electrode potential measured as a function of the ratio of oxidant added to the initial activity of the reductant, that is, E versus $[Ox]/[Re]$.

When $[Ox]/[Re] = 1$, equivalence is reached. Below equivalence, $[Re^+] = [Ox]$ and $[Re^-] = [Re] - [Ox]$, such that

$$\frac{[Re^+]}{[Re^-]} = \frac{[Ox]}{[Re] - [Ox]} = \frac{[Ox]/[Re]}{1 - [Ox]/[Re]} \qquad (5.4)$$

Beyond equivalence, $[Ox^-] = [Re]$, and $[Ox^+] = [Ox] - [Re]$, yielding

$$\frac{[Ox^+]}{[Ox^-]} = \frac{[Ox] - [Re]}{[Re]} = \frac{[Ox]}{[Re]} - 1 \qquad (5.5)$$

Solution of (5.4) and (5.5) for selected increments of $[Ox]/[Re]$ from 0 to 2.0 is given in Table 5.2. The potential is calculated using a 30-mV/decade change in activity ratio. The titration curve is plotted in Fig. 5.1

Observe that the titration curve is not symmetrical. The accumulation of spent oxidant Ox^- reduces the sensitivity of the electrode to excess oxidant Ox^+. Thus the relatively high-gain region below E_{Re} in Fig. 5.1 is not easily reproduced at the other end of the scale —it would require a several-fold excess of oxidant. This is typical of

Table 5.2 Solution of (5.4) and (5.5)

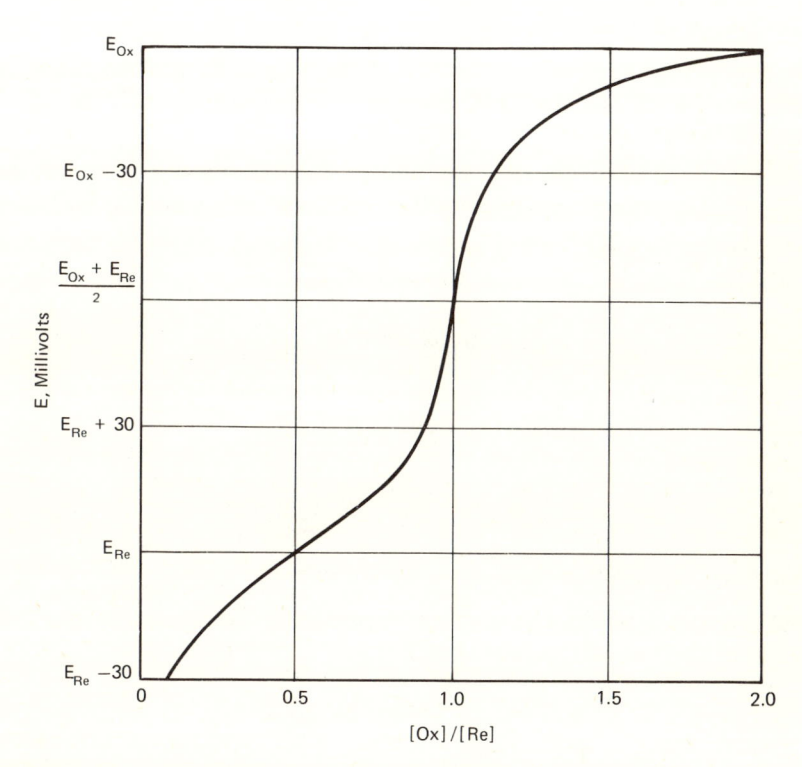

[Ox]/[Re]	[Re$^+$]/[Re$^-$]	[Ox$^+$]/[Ox$^-$]	E
0	0	—	$-\infty$
0.1	0.11	—	$E_{Re} - 29$
0.2	0.25	—	$E_{Re} - 18$
0.5	1.0	—	E_{Re}
0.8	4.0	—	$E_{Re} + 18$
0.9	9.0	—	$E_{Re} + 29$
1.0	—	0	$0.5 E_{Re} + 0.5 E_{Ox}$
1.1	—	0.1	$E_{Ox} - 30$
1.2	—	0.2	$E_{Ox} - 21$
1.5	—	0.5	$E_{Ox} - 9$
1.8	—	0.8	$E_{Ox} - 3$
2.0	—	1.0	E_{Ox}

FIGURE 5.1. A titration curve for a combined oxidation-reduction reaction, where each reactant exchanges two electrons with the other.

situations where an oxidant is manipulated to control the concentration of some reductant. In the opposite case where reductant is manipulated for control, increasing flow will lower the potential, the sensitivity decreasing with addition of excess reductant.

It is difficult to generalize beyond the foregoing, since there is such a great difference from one reaction to the next. In many reactions, more than one species takes part, as Table 5.1 indicates. All the ions appearing on each side of the reaction equation must also appear in the Nernst equation. For the reaction,

$$ClO^- + H_2O + 2e \rightarrow Cl^- + 2OH^-$$

the Nernst equation is

$$E = 890 + \frac{2.3RT}{2F} \log \frac{[ClO^-]}{[Cl^-][OH^-]^2} \tag{5.6}$$

The potential is thus affected by pH— in fact twice as sensitive to pH as to [ClO$^-$]. As above, acids and bases take part in many of these reactions, so that pH control is required as well as redox control in virtually every application.

Observe that Table 5.1 includes two reduction equations for oxygen: one involving hydrogen ions and one yielding hydroxyl ions, each with its own E_0. For the reduction in the presence of hydrogen ions, the potential developed is

$$E = 1229 + \frac{2.3RT}{4F} \log [H^+]^4$$

For the reduction generating hydroxyl ions:

$$E = 401 + \frac{2.3RT}{4F} \log \frac{1}{[OH^-]^4}$$

The actual potential developed in both cases is identical, E_0 only being dependent on the chosen expression. To illustrate, the two equations above will be set equal. Then, at 25°C,

$$828 = -59.16 \log [H^+][OH^-]$$

The product [H$^+$][OH$^-$] at 25°C is 10^{-14}; multiplication of 59.16 by the exponent -14 is 828, so we have an identity.

OXIDATION OF CYANIDE IONS

To illustrate the relationship between the reaction and the resulting potential measurement, two very common examples will be used. The first redox reaction to be described is the oxidation of cyanide by chlorine; the second is the reduction of chromates by sulfite ions. The two are the most familiar applications of redox control, often being conducted side by side in the same facility. Both cyanide and chromates are wastes from metal-finishing plants.

Overall Reactions

The destruction of cyanide takes place in two steps: in the first, cyanide is oxidized to cyanate in a highly alkaline medium, followed then by further oxidation to carbonate and nitrogen in a more neutral environment. Since cyanate is about 1000 times less toxic than cyanide, the second step of the reaction is often omitted. The overall equation for the oxidation of sodium cyanide to cyanate is

$$Cl_2 + 2NaOH + NaCN \rightarrow NaCNO + 2NaCl + H_2O \quad (5.7)$$

Taking into account the molecular weights of the reactants, about 1.4 lb of chlorine and 1.6 lb of caustic are required per pound of sodium cyanide. Destruction of cyanate by further chlorination is described by the following overall equation:

$$3Cl_2 + 6NaOH + 2NaCNO \rightarrow 2NaHCO_3 + N_2 + 6NaCl + 2H_2O$$

$$(5.8)$$

This second step consumes approximately 2.2 lb of chlorine and 2.5 lb of caustic per pound of sodium cyanide.

Sodium cyanide is not the only cyanide salt in metal-plating wastes. In fact, in the plating processes, many of the metals used dissolve to a certain extent in the bath, so that the rinse waters and spent solutions contain cyanides of zinc, cadmium, copper, nickel, silver, and iron. Much of the cyanide may also exist as complex ions, depending on the pH and concentration of both the cyanide and the metal ions. Most can be oxidized by chlorination, however.

108 Reduction-Oxidation

Although the complex ions are not directly oxidized, they are in equilibrium with cyanide ions as shown in (4.17). So as cyanide ions are removed, complex ions will dissociate to take their place. The only complex cyanides which are too stable to be destroyed by chlorination are the iron complexes. The ferrocyanide ion is oxidized by chlorine to ferricyanide, but there the reaction stops. If the ferricyanide ion is in sufficient quantity to warrant removal, it can best be precipitated by adding a ferrous salt.

Regardless of the many ingredients in the plating waste, all of the cyanide which is to be destroyed originally was brought into the plant as sodium cyanide. Therefore the demand for caustic and chlorine can be calculated, knowing the original sodium cyanide concentration in given baths and rinse solutions.

Hydrolysis of Chlorine

Each of the reactions given above is much more involved than the equations suggest. The first depends on the hydrolysis of chlorine to the hypochlorite ion, oxidation of the cyanide ion to cyanogen chloride, and subsequent hydrolysis to the cyanate and chloride ions. Chlorine can oxidize in the gaseous state as oxygen does, but it does not exist as Cl_2 in aqueous solutions. Instead, the diatomic molecule splits into ions wherein it exhibits valence states of $1+$ (the oxidant) and $1-$ (reduced):

$$Cl_2 + 2OH^- \rightarrow ClO^- + Cl^- + H_2O \tag{5.9}$$

Equation (5.9) is not an oxidation-reduction reaction. Only half of the chlorine actually takes part in the subsequent oxidation:

$$ClO^- + 2H_2O + 2e \rightarrow Cl^- + 2OH^- \tag{5.10}$$

These two electrons are donated by the cyanide.

The Nernst equation for the reduction of the hypochlorite ion was given in (5.6). Observe that the maximum ratio of $[ClO^-]/[Cl^-]$ is 1.0, as a result of the hydrolysis. If the reaction were conducted at pH 9, $[OH^-]^2$ would be 10^{-10}. Solving (5.6) for these conditions yields

$$E = 890 + 30 \log 10^{10} = 1190 \text{ mV}$$

Subtracting approximately 200 mV for a 4 M Ag, AgCl reference electrode yields

$$E - E_{ref} = 990 \text{ mV}$$

At pH 10, this would decrease by 60 mV to 930. As hypochlorite is consumed, this potential will fall.

The ClO^- is the active group for the oxidation of cyanide. If the pH of the hypochlorite solution is too low, however, the weak acid HOCl will form, which, in its undissociated state, does not react:

$$H^+ + ClO^- \rightleftharpoons HOCl \qquad (5.11)$$

The pK of HOCl is 7.5; at pH values below 7.5 more than 50% of the ClO^- ions are tied up as HOCl, whereas at pH 8.5 and above, less than 10% are undissociated.

Formation and Hydrolysis of CNCl

Reference 2 gives the oxidation of cyanide as producing the intermediate compound cyanogen chloride:

$$CN^- + H_2O + ClO^- \rightarrow CNCl + 2OH^- \qquad (5.12)$$

This mechanism has been established by conducting the oxidation at a pH of 7 or less, thus inhibiting the hydrolysis of CNCl to CNO^-. Cyanogen chloride is as toxic as the cyanide ion, therefore the reaction must be conducted above pH 7 for hydrolysis to proceed:

$$CNCl + 2OH^- \rightarrow CNO^- + Cl^- + H_2O \qquad (5.13)$$

The hydrolysis simply consumes the hydroxyl ions formed in the oxidation, so no additional amount of caustic is required, above that used in the hydrolysis of the chlorine.

Reduction-Oxidation

Reference 2 cites the actual oxidation as being instantaneous, while the hydrolysis is both time and pH dependent. A decrease in CNCl content from 10 to 0.1 ppm is reported in 3 min at pH 9, in 9 min at pH 8, and incomplete (i.e., equilibrium is established) at pH 7.1 and below.

A standard potential is available in Table 5.1 for the combined oxidation and hydrolysis:

$$CN^- + 2OH^- - 2e \rightarrow CNO^- + H_2O \tag{5.14}$$

Equation (5.14) is an oxidation equation as opposed to the reduction equation given in Table 5.1. The Nernst equation is

$$E = -970 + \frac{2.3RT}{2F} \log \frac{[CNO^-]}{[CN^-][OH^-]^2} \tag{5.15}$$

If half of the cyanide were oxidized to cyanate at pH 9, the potential would be

$$E = -970 + 30 \log 10^{10} = -670 \text{ mV}$$

Again, subtracting 200 mV for the reference electrode yields

$$E - E_{ref} = -870 \text{ mV}$$

Notice how far apart the calculated potentials for the two reactions are. This great difference creates a very sensitive titration curve, and minimizes the effect of pH variations. Changes in pH of one unit shift both potentials by the same 60 mV, while their difference remains at 1860.

Actual oxidation of cyanide by chlorine under potential control does not give the ideal titration curve of the form shown in Fig. 5.1, however. An excess of chlorine raises the potential to an intermediate plateau at about +500 to 600 mV, as shown in Fig. 5.2. At this point a second stage of oxidation begins, consuming more ClO^- ions to stop the rise in potential.

Oxidation of Cyanate Ions

An excess of chlorine above the 1:1 mol ratio with cyanide can oxidize the cyanate to nitrogen and bicarbonate. The standard

Oxidation of Cyanide 111

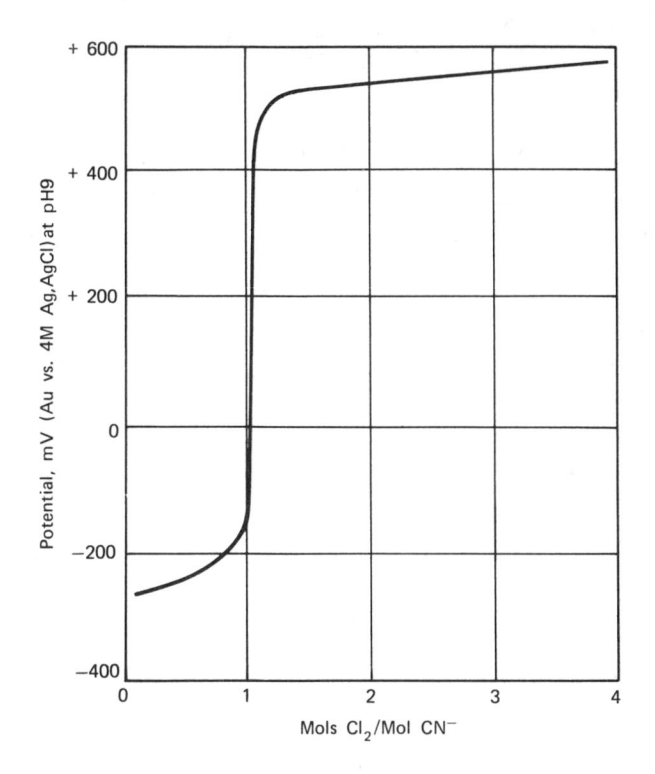

FIGURE 5.2. Titration of chlorine against cyanide as measured with a gold electrode at pH 9.

potential for the reaction

$$2CNO^- + 4OH^- - 6e \rightarrow N_2 + 2HCO_3^- + 2H^+ \quad (5.16)$$

does not appear in any tables, however.

According to Reference 2, this reaction depends on HOCl for an oxidant, and therefore is favored by pH values below 8.5 where HOCl formation is enhanced. Retention time required for complete destruction of cyanate is reported as 7.5 min at pH 7.5, 10 min at pH 8.4, 15 min at pH 9.2, and 80 min at pH 9.9. Here again, pH control is essential to stable control of reduction-oxidation potential.

112 Reduction-Oxidation

Following oxidation, the waste pH is adjusted if necessary, and the metal ions present are allowed to precipitate as hydroxides.

The Plant and Its Controls

Figure 5.3 shows how the process might be arranged for control of cyanide oxidation. The flow of chlorine is typically controlled by a servo driven valve in a chlorinator, such as that supplied by Wallace and Tiernan, Inc.* Water is used as a vehicle to transfer the chlorine into the reaction vessel with an eductor. Caustic is added to the water upstream of the eductor to assist in hydrolysis of the chlorine.

Control of both pH and solution potential is required. Caustic should be set in ratio to the chlorine delivered to the system, otherwise variations in delivery will upset pH. To achieve this, the output

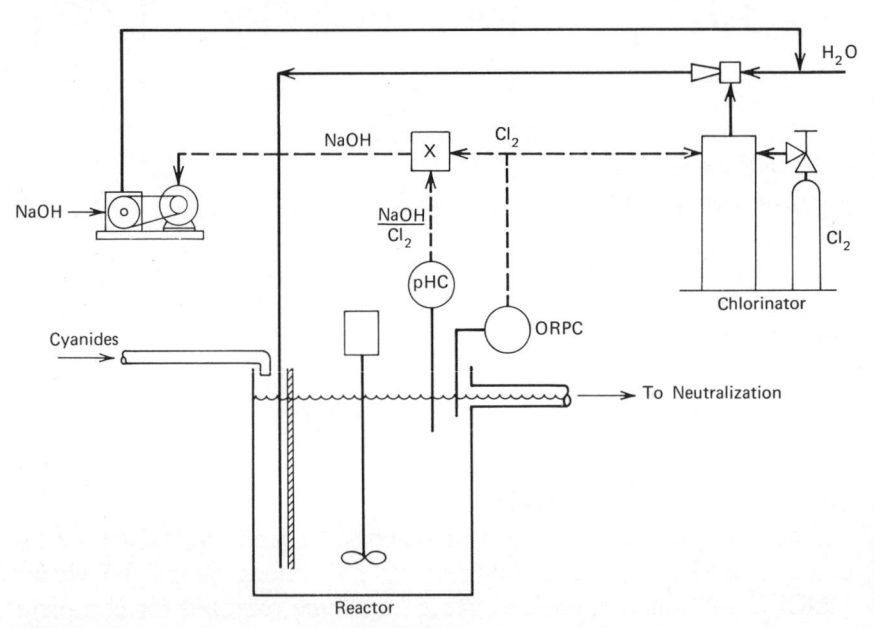

FIGURE 5.3. In order to control both redox potential and pH, chlorine and caustic must be added in a controlled ratio.

*Wallace and Tiernan, Inc., Industrial Products Division, 25 Main St., Belleville, N.J. 07109.

of the potential controller sets chlorine flow directly, and the caustic flow through multiplication by the output of the pH controller. The latter signal is therefore the caustic-to-chlorine ratio, adjusted automatically to control pH.

Both the hydrolysis of cyanogen chloride and oxidation of cyanate are time dependent, which affects the response of the redox control loop. A time constant of 2 to 3 min has been reported for the response of the platinum electrode to a change in chlorine flow, compared to 1 min for a response to a change in cyanide flow.[3] The difference may be due to an insensitivity to the formation of CNCl.

For oxidation to cyanate alone, the control point for the redox potential may be safely set at about +300 mV relative to the 4 M Ag, AgCl reference electrode. The titration curve is quite nonlinear and not symmetrical about this point. A nonlinear controller (see Chapter 8) is recommended, modified for high gain on one side of the set point only. Even with these precautions, this control loop is highly variable, and due to its great sensitivity, some degree of oscillation will normally be encountered. Figure 5.4 is a typical record of pH and redox control on a cyanide oxidation system using the controller just mentioned.

If complete destruction of the cyanide is required, the two stages may be conducted sequentially in a single tank, if batch operation is possible. Where the plant must be operated continuously, two complete stages shown in Fig. 5.2 are recommended, each with 10 to 15 min residence time. For optimum results, the first stage should be controlled at about pH 9 with the second at pH 7.5. Although both reactions could theoretically be conducted together in a single stage at pH 8.5, efforts to do this have not been generally successful.

REDUCTION OF CHROMATE IONS

Chromates are used in metal finishing as corrosion inhibitors, either as the chromate CrO_4^{2-} or dichromate $Cr_2O_7^{2-}$ ions. In these ions, chromium appears with a valence of 6+, in which condition it is both toxic and soluble. Reduction to a valence state of 3+ allows precipitation as chromium hydroxide. Several different reducing agents are in common use, each with its own characteristics; each will be touched upon.

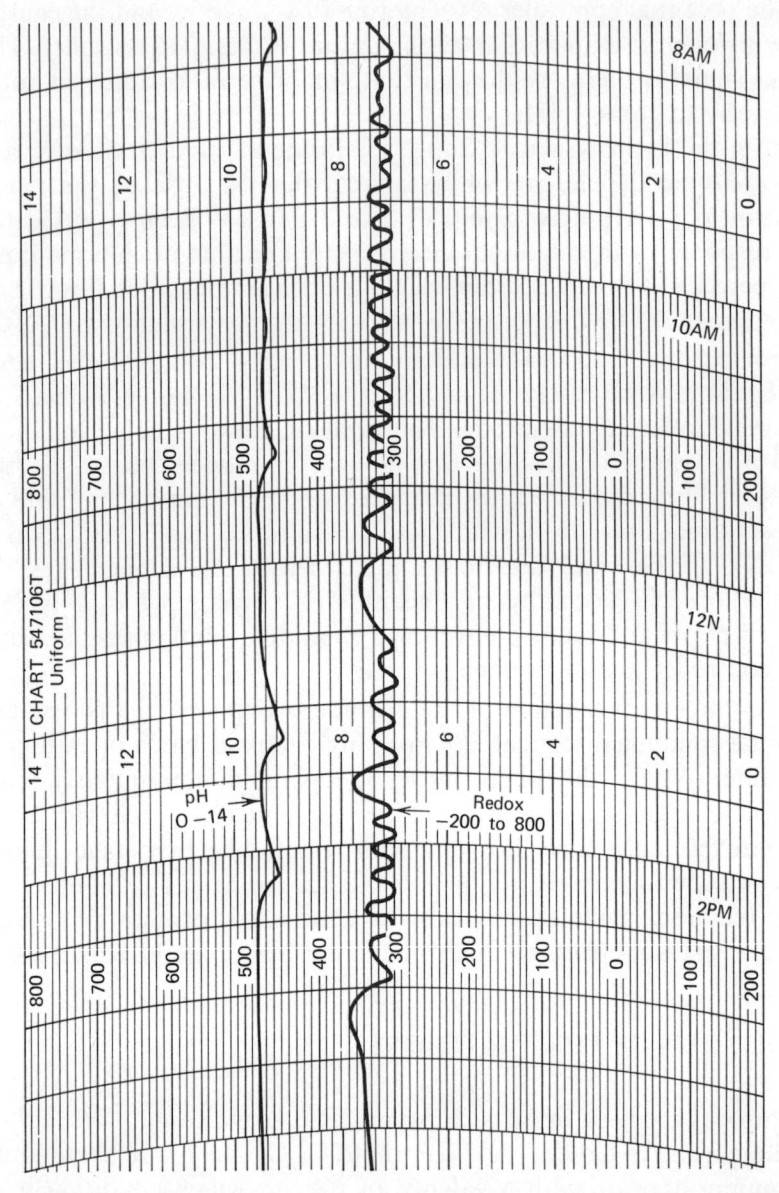

FIGURE 5.4. Control over redox potential illustrated by this plant record is quite acceptable, considering the extreme sensitivity and nonlinearity of the measurement.

114

Overall Reactions

Although the actual state of chromium in the waste at pH 2 or thereabouts is certain to be as chromic acid, H_2CrO_4, it is introduced to the plant ordinarily as sodium dichromate, $Na_2Cr_2O_7$; consequently, the overall reactions below will be written as reductions of dichromate, to allow estimation of reagent requirements per pound of dichromate entering the plant. Several reducing agents are available: metallic iron, ferrous sulfate, and various forms of sulfites. Since the last class of agents is the most popular, equations will be written for each of them.

Sulfur dioxide may be introduced in the gaseous form as the reducing agent, in which case the reaction proceeds as follows:

$$Na_2Cr_2O_7 + 3SO_2 + H_2SO_4 \rightarrow Cr_2(SO_4)_3 + Na_2SO_4 + H_2O \quad (5.17)$$

Three sulfites are available as reducing agents, all of which are crystalline solids that must be dissolved and fed as a solution. Their sulfur dioxide and acid content vary from one to another. The overall reaction with sodium sulfite is

$$Na_2Cr_2O_7 + 3Na_2SO_3 + 4H_2SO_4 \rightarrow Cr_2(SO_4)_3 + 4Na_2SO_4 + 4H_2O$$
$$(5.18)$$

Sodium bisulfite requires less acid:

$$Na_2Cr_2O_7 + 3NaHSO_3 + 2.5H_2SO_4 \rightarrow Cr_2(SO_4)_3 + 4H_2O + 2.5Na_2SO_4$$
$$(5.19)$$

Sodium metabisulfite is essentially an anhydrate of sodium bisulfite:

$$Na_2S_2O_5 + H_2O \rightarrow 2NaHSO_3 \quad (5.20)$$

Its overall reaction with dichromate is

$$Na_2Cr_2O_7 + 1.5Na_2S_2O_5 + 2.5H_2SO_4$$
$$\rightarrow Cr_2(SO_4)_3 + 4H_2O + 2.5Na_2SO_4 \quad (5.21)$$

116   Reduction-Oxidation

The stoichiometric requirements for each of these reagents per unit of $Na_2Cr_2O_7$ are summarized in Table 5.3. Some losses will be experienced as well as air oxidation of SO_2, so actual usage can be expected to exceed these values somewhat.

Table 5.3   Reagent Requirements per Unit Weight $Na_2Cr_2O_7$

Reagent	Weight of Reagent	Weight of H_2SO_4
SO_2	0.732	0.375
Na_2SO_3	1.433	1.500
$NaHSO_3$	1.192	0.935
$Na_2S_2O_5$	0.904	0.935

The Reduction Reaction

Reduction proceeds in an acid medium as follows:

$$Cr_2O_7^{2-} + 14H^+ + 6e \rightarrow 2Cr^{3+} + 7H_2O \tag{5.22}$$

As indicated by the reaction, the potential is highly sensitive to solution pH:

$$E = 1530 + \frac{2.3RT}{6F} \log \frac{[Cr_2O_7^{2-}][H^+]^{14}}{[Cr^{3+}]^2} \tag{5.23}$$

At a pH of 3, $\log [H^+]^{14}$ contributes -520 mV:

$$E = 810 + \frac{2.3RT}{6F} \log \frac{[Cr_2O_7^{2-}]}{[Cr^{3+}]^2}$$

Variations in pH affect E by 140 mV/unit, compared to 10 mV/decade change in $Cr_2O_7^{2-}$ activity and 20 mV/decade change in Cr^{3+} activity. Needless to say, pH control is important.

This reaction is not instantaneous, and, in fact, the residence time required for completion is directly related to pH, as Fig. 5.5 indi-

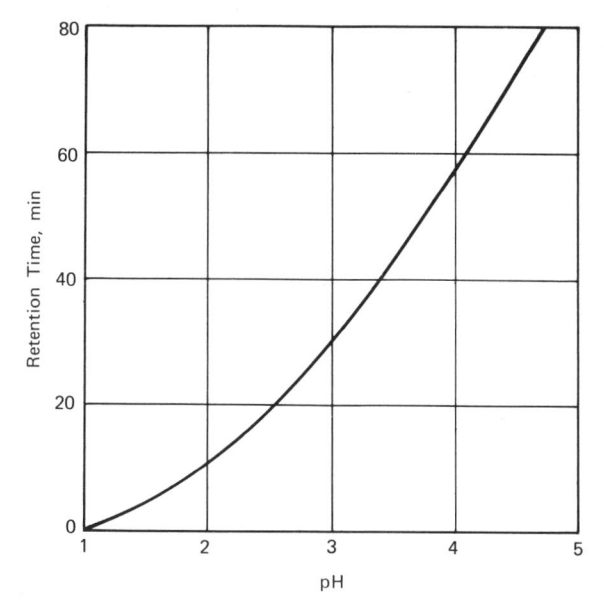

FIGURE 5.5. Retention time required for complete reduction of chromate increases with the pH of the solution.

cates. This is not surprising, since hydrogen ions are consumed in the reaction and thereby affect its forward rate.

Following reduction, the effluent should be neutralized with lime or caustic to a pH of 8 to 9 for effective precipitation of chromium hydroxide. A solubility curve versus pH is given in Fig. 3.6.

The Oxidation Reaction

Metallic iron is an effective reducing agent, although difficult to handle. For small amounts of chromates, passage through a bed of iron turnings at a pH<3 can reduce the chromates with formation of ferric ions. Ferrous sulfate solution has also been used as a reducing agent, where it is plentiful. Its standard reduction potential of +685 mV in acid solution is rather close to the 810 mV estimated at pH 3 for the chromate reduction. Control seems to be unresponsive owing to the insensitivity of the titration curve, and reagent usage tends to be high.

118   Reduction-Oxidation

The preferred reducing agent is the sulfite ion, which is oxidized in the following manner:

$$SO_3^{2-} + H_2O - 2e \rightarrow SO_4^{2-} + 2H^+ \quad (5.24)$$

Its reduction potential is

$$E = 170 + \frac{2.3RT}{2F} \log \frac{[SO_4^{2-}][H^+]^2}{[SO_3^{2-}]} \quad (5.25)$$

Here again, E changes with pH, but only at the rate of 60 mV/pH unit. At pH 3, and equal activitites of SO_3^{2-} and SO_4^{2-}, E would be about -10 mV. Lowering the pH raises this potential less than the dichromate chromium potential, forcing the two farther apart. This improvement in sensitivity is indicated in the titration curves[3] given in Fig. 5.6.

The titration curves differ markedly from the ideal—this is actually the case with most redox potential titrations. Remember that the electrodes are sensitive to all ions present in some degree or

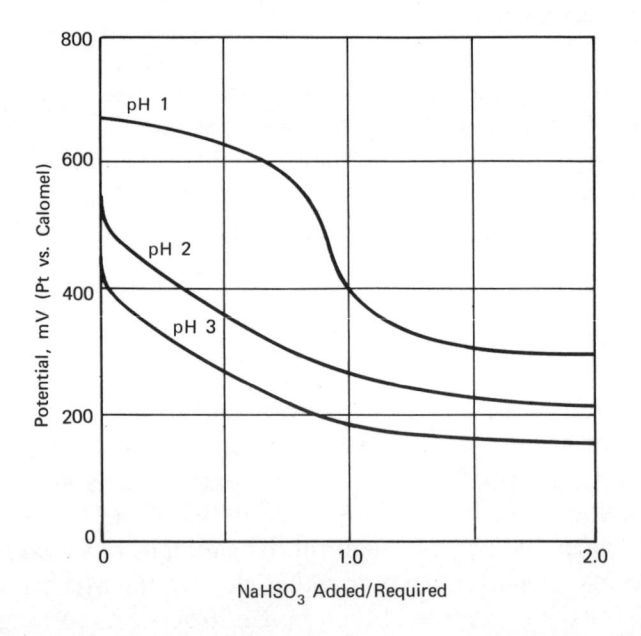

FIGURE 5.6. Titration curves of dichromate with sodium bisulfite are far from ideal, especially at higher pH levels.

another. Oxygen from the air will consume a significant portion of the sulfite ions in open vessels. Even if the reaction is conducted in a closed vessel, the chromate solutions are invariably exposed to air prior to reduction, and so contain some dissolved oxygen. As a result, the O_2–O^{2-} couple does affect the measured potential to some degree.

The Control System

Control over the reduction reaction simply requires pH and ORP control in the reduction vessel followed by pH control in the subsequent neutralization vessel. About +300 mV versus the 4 M Ag,AgCl electrode and pH 2.3 are recommended control points. Variable-speed metering pumps are commonly used for adding sulfuric acid and sulfite solution, while a gas feeding system similar to the chlorinator is used for sulfur dioxide. The control loops may be operated independently of one another, since pH control in the region of 2 to 3 is 300 times less demanding than in the 8.5 to 9.5 region used for chlorination of cyanide. Neutralization with lime or caustic to pH 8 or thereabouts for precipitation is no different for this system than for any acid neutralization, except that a settling tank is required to clarify the treated effluent. Figure 5.7 outlines the system.

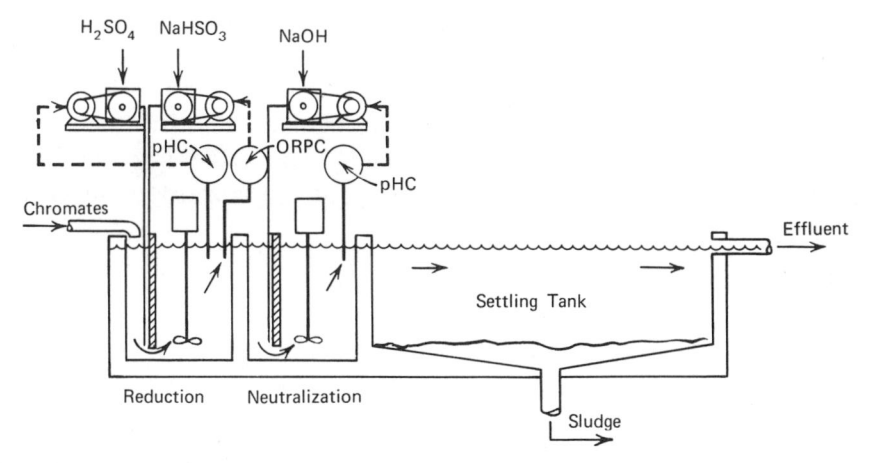

FIGURE 5.7. Neutralization and precipitation are essential parts -f chromium-removal systems.

ELECTRODE SELECTION

Although electrodes in general were discussed in Chapter 1, their selection for reactions involving oxidation and reduction had to be deferred until the reactions themselves were presented. Even for the more common reactions, a choice of noble metal electrodes is presented, while more specific electrodes and combinations thereof deserve examination for special applications.

Noble Metal Electrodes

Electrodes are available in billet or foil construction of silver, gold, or platinum. The platinum electrodes may be of the polished metal or "platinized," the latter having a black rough surface containing a much larger area than the smooth metal. Platinum is ordinarily considered an inert metal and therefore the usual choice for redox potential measurements such as in chromate reduction. However, it does catalyze many reactions, and in so doing, may yield a distorted titration curve. It is not recommended for cyanide oxidation, for example, since at potentials below about +200 mV, hydrogen evolution occurs at the electrode surface, yielding a plateau unrelated to the cyanide activity. Figure 5.8 compares this titration curve to that of the gold electrode under the same conditions.[3]

Although gold is soluble in high concentrations of cyanide, this does not constitute a problem, since the control point of +300 mV is far removed from the problem area.

Silver electrodes are more reactive and hence are not often used to measure redox potential. They are particularly sensitive to strong oxidizing agents such as chlorine, being converted into an ion-selective electrode by a surface coating of the corresponding salt.

The Antimony Electrode

Certain metals, such as antimony, molybdenum, and tungsten, react with water to form an oxide with the release of hydrogen ions:

$$2Sb + 3H_2O \rightleftharpoons Sb_2O_3 + 6H^+ + 6e^- \qquad (5.26)$$

This is essentially a reduction-oxidation reaction with a strong influence from the hydrogen ion. The potential developed by the antimony electrode in the presence of hydrogen ions is

$$E = 145 + \frac{2.3RT}{F} \log[\text{H}^+] \qquad (5.27)$$

The principal use of the antimony electrode is to measure pH. It was not discussed under electrodes in Chapter 1, since its peculiar sensitivity to hydrogen-ion activity is a result of oxidation at the electrode surface. Its advantage is a low impedance like other redox

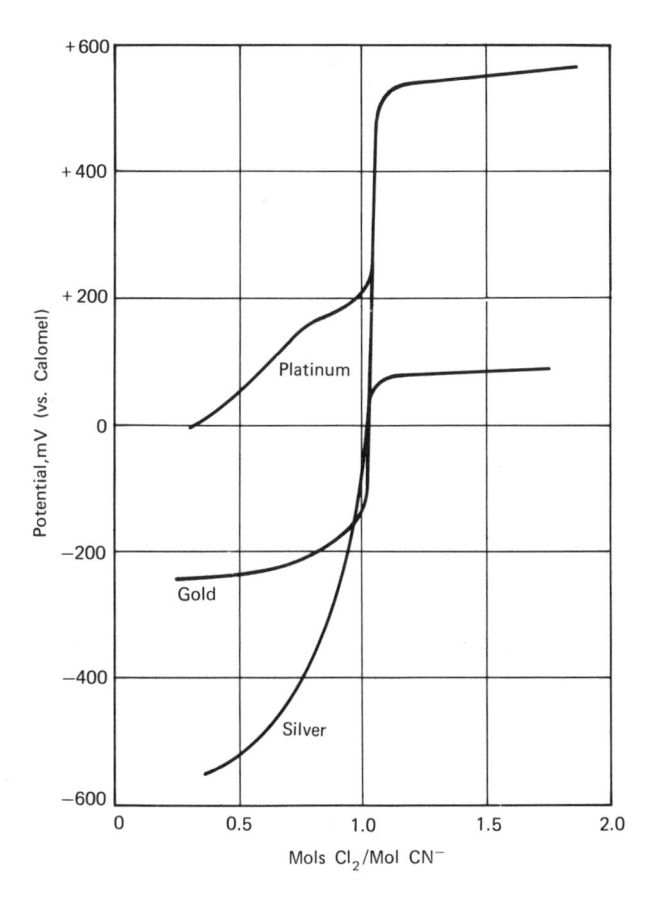

FIGURE 5.8. The gold electrode gives the most representative potential over the widest range for cyanide oxidation.

electrodes, in the order of 10,000Ω. However, it is affected by the presence of oxidizing and reducing agents in the solution.

Its response to the logarithm of hydrogen-ion activity is only linear in the middle range of the pH scale, from about 3 to 8, depending on what other ions are in the solution. The electrode is quite temperature sensitive, with a coefficient varying from 1 to 3 mV/°C, increasing with pH, as contrasted to about 0.4 mV/°C for the glass pH electrode. In addition, it has no symmetry with commonly used reference electrodes, making temperature compensation both difficult and necessary if any degree of accuracy is expected.

Calibration can be a real problem. The buffer solution should contain the same levels of oxidizing and/or reducing agents as the process solution, otherwise a correct calibration cannot be achieved. The antimony electrode seems to be favored in mineral-extraction processes more than anywhere else, probably due to its low impedance, ruggedness, and the absence of interfering agents in these solutions

Special Reference Electrodes

Some reactions that appear amenable to redox-potential measurement are not. One of this class is the formation of hypochlorite bleach by the chlorination of caustic:

$$2NaOH + Cl_2 \rightarrow NaClO + NaCl + H_2O \qquad (5.28)$$

An equilibrium is established between the chloride and hypochlorite ions given in (5.10), with the associated potential described by (5.6). Notice that a change in the ratio of chlorine to caustic only affects the redox potential through an adjustment of hydroxyl-ion activity, since the $[ClO^-]/[Cl^-]$ ratio is unity. But this is because (5.28) simply a hydrolysis, not a reduction or oxidation. A pH measurement may indicate the state of the reaction as readily.

If a chloride-ion electrode is used as a reference, the effect of that ion on the potential difference is altered. Restating (5.6) for the noble metal electrode:

$$E = 890 + \frac{2.3RT}{2F}(\log[ClO^-] - \log[Cl^-] - 2\log[OH^-])$$

The voltage developed for the chloride-ion electrode from (1.16) is

$$E_{ref} = 236 - \frac{2.3RT}{F} \log [Cl^-]$$

The difference between the two voltages is

$$E - E_{ref} = 654 - \frac{2.3RT}{2F} (\log[ClO^-] + \log[Cl^-] - 2\log[OH^-])$$

Note that the logarithm of the chloride-ion activity has not been eliminated from the expression, but is now added rather than subtracted. Since chlorine generates hypochlorite and chloride ions on a one-for-one basis, their activities should be identical. Then $[ClO^-]$ can be substituted for $[Cl^-]$:

$$E - E_{ref} = 654 - \frac{2.3RT}{F} \log \frac{[ClO^-]}{[OH^-]} \qquad (5.29)$$

If pH is controlled by adding caustic, hypochlorite-ion activity can be controlled as the difference between redox potential and chloride-ion activity, as given in (5.29).

Reference 4 cites a custom-made reference electrode, outwardly appearing like a flowing silver chloride or calomel electrode. Within, however, is a noble metal wire identical to the measuring electrode, surrounded by a standard solution with a liquid junction to the process. The potential difference developed between the two electrodes then represents the difference in ion activity between the process and the standard.

The standard solution should be prepared to match the process solution at the desired control point. Then both electrodes will develop the same potential when the process is at the control point, and errors due to liquid-junction potential and temperature will be eliminated.

The standard solution must obviously be stable with time and temperature, free of suspended solids, and easy to handle. If the process solution does not have all these properties, it may be necessary to prepare a standard which simulates the process in voltage and liquid-junction potential, but without the other undesirable features. The reader is encouraged to use his imagination in applying this technique.

REFERENCES

1. R. C. Weast, *Handbook of Chemistry and Physics*, 50th ed., The Chemical Rubber Company, Cleveland, Ohio, 1969.
2. N. S. Chamberlin and H B. Snyder, Jr., *Technology of Treating Plating Wastes*, Wallace and Tiernan, Inc., Belleville, New Jersey.
3. G. Mattock, Automatic Control in Effluent Treatment, *Transactions of the Society of Instrument Techniques*, December 1964.
4. W. N. Greer, Measurement and Automatic Control of Etching Strengths of Ferric Chloride, *Plating*, October 1961.

PART 3
CONTROL

6

THE FUNDAMENTALS OF COMPOSITION CONTROL

The ultimate service of any measuring device is in a control system, for without control, measurement is little more than speculation. Control demands much more than measurement technology alone. Many parts go into the making of a control loop: Controller, valve, piping, vessels, agitation, and the reaction itself all share in the success or failure of an installation. But the first step in combining these elements into an effectively functioning system is to understand the principles of feedback control.

FEEDBACK CONTROL

The basic means for achieving regulation over any process variable is negative feedback. A measurement made of the state of the process is acted on by the controller to change that state, if it is undesirable. Then the effect of that control action is measured and sent to the controller to modify further action. The flow of information from the process to the controller back to the process is negative in sense, in that the action taken is in the direction that will reduce an error rather than augment it.

With feedback comes the prospect of instability, however: Too much control action can overcorrect an error, causing another error in the opposite direction. Thus periodic oscillations readily develop which are characteristic of negative feedback. To achieve effective control requires corrective action of such a degree that disturbances are rapidly countered, yet not to the degree that the controller disturbs itself into sustained oscillations. Therefore the science of feedback control is directed toward traveling that narrow path between ineffectiveness and instability.

Because the path is narrow, systems which are designed to exhibit a certain small margin of stability will approach optimum effectiveness. The benchmark of feedback control effectiveness which is used throughout this presentation will then be the limit of stability.

Oscillation through Phase Shift

For simplicity, consider a feedback loop consisting only of a controller and a process, as shown in Fig. 6.1. A measurement of the stream composition, the controlled variable c, is compared with the set point r. Their difference, the deviation or error e, is the signal acted on by the controller. The controller's output is the manipulated variable m, which in the case of composition control is the flow of reagent to the process. This manipulated reagent flow acts in opposition to the uncontrolled load also entering the process. In a neutralization system, the load might be acid which must be neutralized by manipulating the proper flow of basic reagent to the process.

Feedback is negative since an increasing concentration calls for a

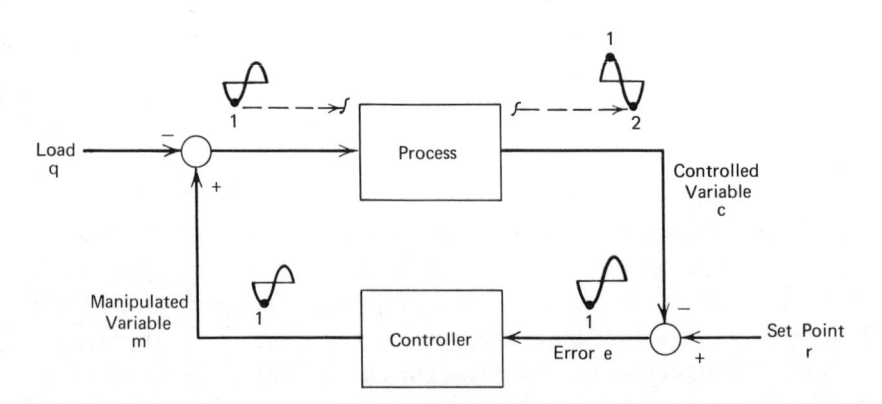

FIGURE 6.1. Allocation of phase shift in a typical feedback loop.

reduction in the manipulated flow of that reagent. A negative sign must appear somewhere in the feedback loop—in Fig. 6.1 it appears at the junction where the error is calculated. In another case, it might appear where the manipulated variable and load come together, indicating that the controlled variable responds positively to the load, as in the case where the load was basic and acid flow was manipulated for pH control.

The most characteristic property of negative-feedback loops is their tendency toward oscillation. In correcting for some deviation of the controlled variable from the set point, it is easy for the controller to overact, developing an opposite deviation at a later time. This, in turn, induces corrective action in the reverse direction, and a continuing cycle results.

Any oscillation has two observable properties which are directly related to the combination of elements in the feedback loop:

1. The *period*, that is, the time between successive cycles, is related to the phase lags in the loop.

2. The *damping* is a function of the attenuation or amplification which these elements provide.

An oscillation of uniform amplitude (i.e., undamped) undergoes a complete 360° shift in phase in passing through the entire loop. Follow the sine waves superimposed on the block diagram in Fig. 6.1 to see where the phase shifting takes place. Start with the controlled variable passing through its minimum value in the first half-cycle. At the summing junction of the controller, it is subtracted from the fixed set point, generating a deviation which is then passing through its positive half-cycle. The negative sign of the summing junction has then effectively introduced 180° of phase shift.

If the controller has no phase lead or lag, the output of the controller will be in phase with the input although its amplitude may differ, depending on the gain set into it. The controller in Fig. 6.1 exhibits no phase shift but some attenuation, judging from the relationship between the input and output sine waves.

The block diagram of Fig. 6.1 shows the manipulated variable entering the process through a positive sign, as in pH responding to the opening of a caustic valve. Consequently, the minimum point in the manipulated-variable cycle must produce a minimum point in the controlled-variable cycle. In order for this to happen, the process must *delay* the effect of m on c for one half-cycle, that is, 180°. For oscillations to be sustained, then, $-180°$ of dynamic phase shift must accompany the 180° developed by the negative sign in the loop, yielding a total of 360°. If the dynamic phase shift in the loop is not $-180°$ at a particular period, the period will change until the requisite $-180°$ is developed.

If the controller should produce $-30°$ of phase shift, however, only $-150°$ would take place across the process, since the total dynamic phase of the loop must be $-180°$. It is also possible for the controller to exhibit a phase lead or positive shift which would allow that much more than 180° lag through the process. When the controller's phase shift is zero, the period of oscillation of the loop is known as the *natural period*.

The natural period of each process is determined by the response of the dynamic elements between the output and input of the controller, that is, valve, tank, mixer, electrodes, and so on. Since the rate of recovery from a disturbance is limited by the natural period, an effort will be made in plant design to improve the dynamic response of the process wherever possible.

Loop Gain

In addition to a dynamic phase lag of 180, a loop gain of unity is necessary to sustain oscillation. If the corrective action taken produces a subsequent error smaller than that which initiated the action, attenuation exists in the loop and damping will ensue. If no attenuation exists, the cycle is perpetuated. Any one element may attenuate, but if another amplifies so as to restore the original signal level, the oscillation will be sustained. Therefore the product of the gains of all the elements in the loop determines its stability.

Gain is here defined as the change in the output of an element divided by the change in the input which caused it. If the product of the gains of all the elements in a loop (loop gain) at the natural period is less than 1.0, damping will result; whereas if it exceeds 1.0, the oscillation will expand. Figure 6.2 illustrates the effect of a loop gain of 0.5. Since, in this example, the excursion in m is one-half that of c at the same instant, the controller gain is 0.5, while the process gain is 1.0. However, any combination yielding a product of

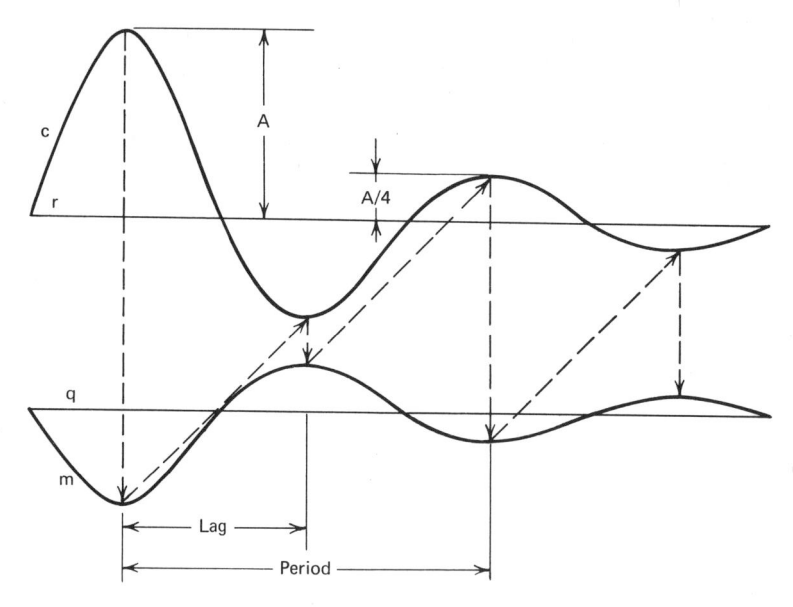

FIGURE 6.2. Relationships between the variables in a quarter-amplitude-damped oscillation.

0.5 will produce the same damping.

Observe that each half-cycle is reduced by one-half, and each full cycle is reduced by one-fourth. This *quarter-amplitude damping*, produced by a loop gain of one-half, is considered optimum for many applications in industry.

The arrows in Fig. 6.2 indicate the sequence of events. A certain value of the controlled variable produces a corresponding output m at the same instant, while a value of m produces a corresponding value of c one half-cycle later.

The gain of the controller is adjustable, while that of the process is not. To obtain the most effective control action, the controller gain should be increased just short of the limit of stability, that is, until the loop gain lies in the vicinity of 0.5. But the actual value of the controller gain at this point depends on the process gain at its natural period. If the process gain is low, the controller gain can be high, requiring only a small error to elicit a given change in the manipulated variable. A low process gain also indicates a low sensitivity to uncontrolled upsets in load. One of the purposes of this and the following chapter is to establish design guidelines

132 Composition Control

directed toward minimizing the process gain.

One objective of great importance in the operation of any feedback control system is a constancy of loop gain. If the gain of one of the elements varies with flow, valve position, composition of the process stream, or the level of the controlled variable, loop gain will change unless suitable compensation is applied. Since instability is generally intolerable, the operator must adjust the controller gain for stability at the highest gain product of the other elements. Then, as the gain of one or another element decreases due to changing conditions, control is less effective than it could be. Obviously it is impractical for an operator to adjust the gain of his controller continuously as conditions change. So to achieve effective control under all conditions, the sources of gain variations must be found and corrected, or compensated by suitable devices in those cases where correction cannot be made.

DYNAMIC GAIN AND PHASE

The dynamic elements in the process—dead time and capacity—produce the phase shift which determines the natural period of oscillation. To predict the speed of response of the process and its control system, the phase characteristics of these elements are of particular importance. But dynamic gain is important, too. Capacity has the property of attenuating a periodic input and therefore impedes rapid changes in both control action and load upsets. A thorough understanding of the dynamic properties of these elements will serve as a foundation for the design of a controllable process.

Dead Time

Dead time[1] is defined as the time interval between the initiation of an action and the first observation of a result. It is caused by transportation of material from the point of manipulation to the point of detection, and is often referred to as "transport delay." Every composition-control loop will contain dead time, since the ions or molecules which are sensed by the measuring device must be transported to that point by a flowing stream. The motive forces in a plant (pumps, agitators, etc.) and the conduits (pipes, channels, and vessels) determine the dead time which the control system will encounter.

Pure dead time is exemplified by the playback of a recording—what is put on the record comes off one dead time later. There is no attenuation or filtering—simply delay. If a sine wave is delayed, its output is shifted in phase from its input by that portion of the cycle which is equal to the dead time:

$$\phi_d = \frac{-360\tau_d}{\tau_o} \tag{6.1}$$

Here the phase shift ϕ_d is expressed in degrees, while dead time τ_d and the period of the wave τ_o are in the same units of time. The negative sign indicates that the output of the element lags the input.

Since dead time does not change the magnitude or form of the signal, its gain is unity, and may be left out of any gain-product calculation.

In processes where neutralization is conducted in a pipe or in-line mixer such as that shown in Fig. 3.3, dead time is the dominant dynamic element. In these cases, the process is most difficult to control since dead time provides no attenuation. With essentially no other phase-shifting elements in the loop, the dead-time element develops the full $-180°$:

$$\phi_d = -180$$

Then (6.1) may be used to find the natural period τ_o:

$$\tau_o = \frac{-360}{-180}\tau_d = 2\tau_d$$

Capacity

Every physical system has capacity to store material or energy. Inertia, heat capacity, gas and liquid volumes, and electrical capacitance all have this capability. A capacitive system may be readily described by a first-order differential equation. As an example, consider a flow of concentrated reagent X being added continuously to a solvent flowing at rate F through a perfectly mixed vessel of constant volume V to form a solution of concentra-

134 Composition Control

tion x. The input of reagent equals the output plus the rate of change of reagent stored in the vessel:

$$X = Fx + V \frac{dx}{dt} \tag{6.2}$$

The concentration of reagent in solution leaving the vessel, if the vessel is perfectly mixed, is also the concentration everywhere within.

Dividing both sides of (6.2) by F allows calculation of the steady-state gain of x with respect to X:

$$\frac{X}{F} = x + \frac{V}{F} \frac{dx}{dt} \tag{6.3}$$

$$\lim_{t \to \infty} \frac{dx}{dX} = \frac{1}{F} \tag{6.4}$$

The coefficient V/F in (6.3) is the time constant of the vessel; its units are those of time. For simplicity, let V/F be replaced by the symbol τ_1.

A first-order capacity such as that described above exhibits a step response of the form

$$\Delta x = \frac{\Delta X}{F}(1 - e^{-t/\tau_1}) \tag{6.5}$$

where e is 2.713, the base of natural logarithms. Observe that the initial slope of the step-response curve traced in Fig. 6.3 is $100\%/\tau_1$

If reagent flow X is varied in a sinusoidal manner, resulting concentration x will exhibit an amplitude and phase shift which are

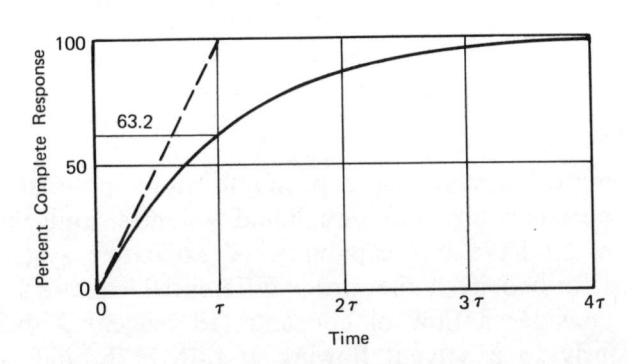

FIGURE 6.3. The step response of a first-order lag.

related both to the time constant of the vessel and the period of the wave[2]:

$$G_1 = \frac{dx(t)}{dX(t)} = \frac{1}{F} \frac{1}{\sqrt{1 + (2\pi\tau_1/\tau_o)^2}} \quad (6.6)$$

$$\phi_1 = -\tan^{-1} \frac{2\pi\tau_1}{\tau_o} \quad (6.7)$$

The terms G_1 and ϕ_1 are defined as the dynamic gain and phase shift of a first-order lag.

Combining Dead Time and Capacity

Neither pure dead time nor pure capacity exists alone in any real process. Vessels are not perfectly mixed, so some time must expire before a change in reagent addition is transported to the sensing element. Consequently, dead time is present in every composition-control system, although usually accompanied by a relatively large capacity.

The period of oscillation of the control loop is determined by the combined phase shift of these two elements. Equations (6.7) and (6.1) should then add to $-180°$. This yields a single equation with one unknown, τ_o, which can be calculated from given or measured values of τ_d and τ_1. Unfortunately, however, a direct solution is unobtainable because τ_o appears in trigonometric form as well as algebraic form. A trial-and-error method is then necessary.

As a first approximation, use the limit of ϕ_1 as τ_o/τ_1 approaches zero:

$$\lim_{\tau_o/\tau_1 \to 0} \phi_1 = -\tan^{-1} \infty = -90° \quad (6.8)$$

Since the total phase lag in the loop is $-180°$, $-90°$ developed by the capacity leaves only $-90°$ for the dead time. The period τ_o may then be calculated from (6.1):

$$\phi_d = -90 = -360 \frac{\tau_d}{\tau_o}$$

$$\tau_o = 4\tau_d \quad (6.9)$$

Composition Control

Then check the validity of the first approximation by substituting for τ_o in (6.7), developing a second approximation for ϕ_1;

$$\phi_1 \approx -\tan^{-1}\frac{2\pi\tau_1}{4\tau_d} = -\tan^{-1}\frac{\pi}{2}\frac{\tau_1}{\tau_d} \tag{6.10}$$

Table 6.1 lists second approximations of ϕ_1 and τ_o calculated from a spectrum of τ_d/τ_1 ratios typical of composition-control systems.

The table indicates that the first approximation, that is, $\phi_1 = -90$ and $\tau_o = 4\tau_d$, is entirely adequate for estimating purposes. As higher τ_d/τ_1 ratios are encountered, control deteriorates; while these unfavorable ratios should be avoided at the outset by proper plant design, they do exist, and are given consideration later in the chapter.

Equation (6.6) describing the dynamic gain of the process can also be simplified by approximation. As ϕ_1 approaches $-90°$, its tangent squared becomes very large with respect to unity. The limit of G_1 under these conditions is

$$\lim_{\phi_1 \to -90} G_1 = \frac{1}{F}\frac{\tau_o}{2\pi\tau_1} \tag{6.11}$$

The gain limit is approached more rapidly than the phase limit due to squaring.

If V/F is substituted for τ_1 in (6.11), the variable F will conveniently disappear from the gain equation:

$$\lim_{\phi_1 \to -90} G_1 = \frac{1}{F}\frac{\tau_o}{2\pi V/F} = \frac{\tau_o}{2\pi V} \tag{6.12}$$

Table 6.1 Approximate Phase and Period

τ_d/τ_1	Tan ϕ_1	ϕ_1 (degrees)	τ_o/τ_d
0.05	31.4	−88.2	3.93
0.10	15.7	−86.3	3.85
0.15	10.5	−84.5	3.77
0.20	7.95	−82.7	3.70
0.25	6.28	−80.9	3.64
0.30	5.23	−79.2	3.58

The remaining term, V, can be considered a constant in most processes, since tank level is usually regulated, either by overflow or by a feedback controller. Dead time, hence the period of oscillation, will be constant if it is primarily determined by agitation or pumping; if dead time varies with flow F, then the period and hence the dynamic gain will vary in direct proportion.

Example 6.1

An acid waste is to be neutralized in a 200-gal agitated vessel by addition of 10% caustic. The dead time in the response of pH to a step change in caustic flow is 10 sec. Estimate the natural period and dynamic gain of the process at a waste flow of 100 gpm:

$$\tau_1 = 200 \text{ gal} / 100 \text{ gpm} = 2 \text{ min} = 120 \text{ sec}$$

$$\frac{\tau_d}{\tau_1} = \frac{10}{120} = 0.083$$

$$\tau_o = 3.9 \tau_d = 39 \text{ sec}$$

$$G_1 = \left(\frac{1}{100 \text{ gpm}}\right)\left(\frac{39}{2\pi 120}\right) = 5.17 \times 10^{-4} / \text{gpm}$$

The dynamic gain term calculated in the foregoing example is obviously incomplete, as its dimension indicates. It is but one term in a series of gains which must be combined to determine the overall response of the process to control action and disturbance.

Reaction Rate Lag

Chemical reactions which are not instantaneous add another dynamic element to the control loop. If the electrode does not sense the manipulated reagent directly but rather a product of its reaction, a delayed response will result. Consider the first-order reaction wherein reagent X forms product C whose concentration c is sensed by the electrode.

138   Composition Control

The rate of the reaction[3] is proportional to the concentration x of the reactant. In a batch reactor this would be described by the equation

$$\frac{dc}{dt} = kx$$

where k is the reaction rate coefficient in units of inverse time. In a continuous backmixed reactor, however, the production of C depends on both the volume of the reacting mixture V and the concentration x:

$$C = Fc + V\frac{dc}{dt} = Vkx \qquad (6.13)$$

The familiar material-balance equation (6.2) must next be modified to include converted product C along with that portion of reactant X which is unconverted:

$$X = F(c + x) + V\left(\frac{dc}{dt} + \frac{dx}{dt}\right) \qquad (6.14)$$

Next, let us solve (6.13) for the unmeasured x in terms of the measured variable c:

$$x = \frac{F}{Vk}c + \frac{1}{k}\frac{dc}{dt}$$

This last equation may be substituted for x in (6.14) and then differentiated and substituted for dx/dt in the same equation. The result is

$$\frac{X}{F} = \left(1 + \frac{F}{Vk}\right)c + \left(\frac{2}{k} + \frac{V}{F}\right)\frac{dc}{dt} + \frac{V}{Fk}\frac{d^2c}{dt^2} \qquad (6.15)$$

Equation (6.15) is a second-order differential equation with real roots. The system it represents is the first-order capacity of time constant $\tau_1 = V/F$ previously described by (6.2) to (6.7), followed by a similar first-order capacity having a time constant $\tau_2 = 2/k$. The properties of the second capacity are identical to the first with the exception of the time constant.

The reaction rate lag adds phase shift ϕ_2 to the loop which increases the period of the control loop in proportion to τ_2:

$$\phi_2 = -\tan^{-1}\frac{2\pi\tau_2}{\tau_o} \tag{6.16}$$

It also introduces another gain term G_2, which is dimensionless and always less than unity:

$$G_2 = \frac{dc(t)}{dx(t)} = \frac{1}{\sqrt{1+(2\pi\tau_2/\tau_o)^2}} \tag{6.17}$$

The effect of the phase lag is considerable in that it can increase the period of oscillation by as much as tenfold, at the same time increasing G_1 proportionally. The reduction in gain offered by G_2 is not as significant, such that the overall effect of τ_2 is an increase in both gain and period.

Since τ_o appears in the equations for all phase angles, its evaluation requires a trial-and-error solution.

STEADY-STATE GAIN

Loop gain consists of the product of the gains of all the elements in the loop—valve, process, electrode, transmitter, and controller. The process gain G_1 was already evaluated in terms of its dynamic response, and was given units of inverse flow. Reaction rate lag G_2 (if present) is dimensionless. The other elements—with the exception of the controller—can generally be assumed to be instantaneous in response, at least in comparison to the process and reaction. Then we only need evaluate the steady-state components of gain for the remaining elements—the valve, electrode, and transmitter.

That portion of the loop seen by the controller has a gain product $G_1(G_2)G_E G_V G_T$ where the subscripts E, V, and T refer to the above-mentioned elements. Each of the gains (except G_2) has units assigned to it, but their product is dimensionless, being controller input divided by controller output. Transmitter gain G_T, for example, is 100% output divided by the input span in pH units, millivolts, or whatever significant units are applied to the electrode potential. The electrode and valve gains are sufficiently variable to warrant more detailed examination.

Electrode Gain

The potential developed by the electrode represents that variable (e.g., pH) by which concentration is indicated and controlled. In the examples given in this text, it will be either pH, pIon, or reduction-oxidation potential (which may be expressed in millivolts). The relationship between the measurement and the amount of reagent added to the stream is the titration curve—the most characteristic mark of any neutralization process.

The gain of a titration curve, from the standpoint of the controller, is the change in pH (or other measurement) from the control point caused by a particular change in reagent concentration. A typical curve is given in Fig. 6.4, to illustrate this concept. The slope and gain are equal only at the control point. For any value of measurement away from the control point, the gain of the curve is the slope of a line drawn from that value through the control point.

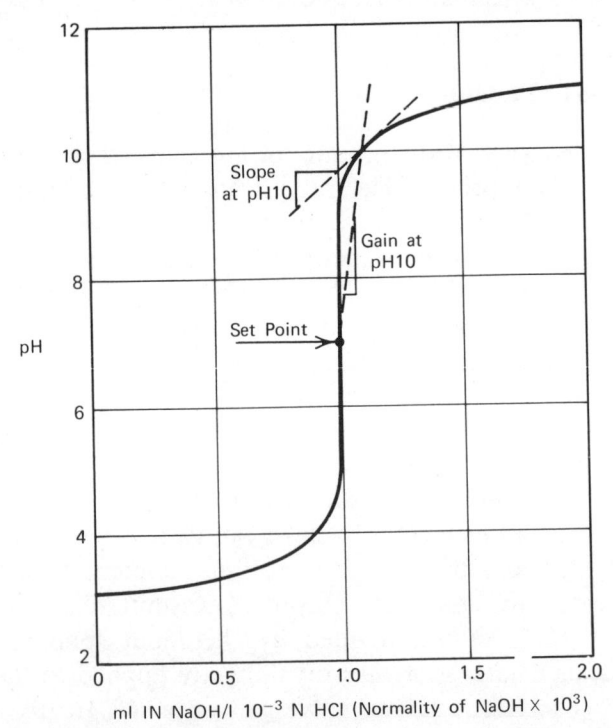

FIGURE 6.4. The gain and slope are equal only at the set point.

The reasoning here is that the controller will make an adjustment to the flow of reagent proportional to the deviation of the measurement from the set point. The slope of the curve at any value of measurement away from the set point has no bearing whatever on action of the controller. If the set point is changed, however, the electrode gain must be reevaluated. Figure 6.5 describes the gain of the titration curve of Fig. 6.4 evaluated at a set point of pH 7.

Most of the processes described in this book have decidedly nonlinear titration curves. As a result, the measurement gain can be expected to be variable. This is the largest stumbling block to effective control of pH and similar variables. If loop gain is to be held constant, some manner of nonlinear compensation must be provided. Gain variations of 2:1 are tolerable, but most titration curves have major gain variations, frequently exceeding 1000:1.

Some curves have such a high gain that stability cannot be provided with conventional controllers (without nonlinear compensation). When this happens, oscillation will expand until an amplitude is reached where (due to decreasing electrode gain) loop gain

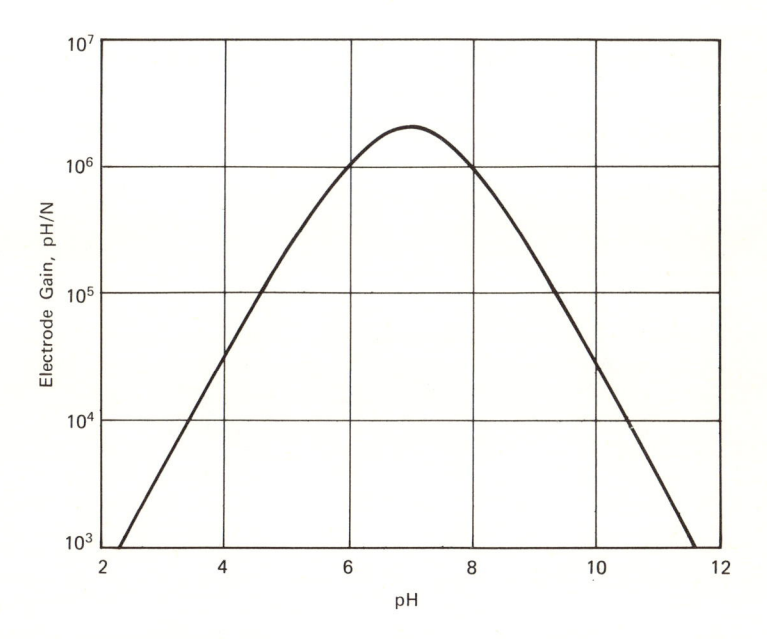

FIGURE 6.5. The gain of the strong acid-strong base titration curve of Fig. 6.4 for a set point of pH 7.

falls to unity. Oscillation will than continue undamped at that amplitude. This is known as a "limit cycle" since it is a uniform oscillation of "limited" amplitude—it is characteristic of loops containing uncompensated nonlinear elements.

Example 6.2

From the strong-acid titration curve of Fig. 6.4, the electrode-gain curve of Fig. 6.5 was plotted. Its maximum value at pH 7 is 2.17×10^6 pH/N which means that 0.46 gal of 1 N NaOH could change 1,000,000 gal from pH 6.5 to 7.5. Multiplying by the dynamic gain from Example 6.1:

$$G_1 G_E = (5.17 \times 10^{-4}/\text{gpm})(2.17 \times 10^6 \text{ pH}/N) = 1120 \text{ pH}/N \text{ gpm}$$

Recall from Chapter 3 that bicarbonate alkalinity up to 10 ppm $CaCO_3$ present in most water supplies can reduce G_E at pH 7 to as low as 1.4×10^4 pH/N. In any given application, the actual value of bicarbonate alkalinity should be determined, in order to estimate G_E properly. If other weak agents or their salts are present or suspected, an actual titration should be made. To arrive at units of pH/N for G_E, the abscissa of the curve should be converted into normality of the titrant (NaOH or HCl) in the titrated solution as was done for Fig. 6.4.

Valve Gain

Valve gain is herein defined as the change in hydrogen or hydroxyl-ion flow from a valve divided by the forcing percentage change in stem position. This definition allows reagent strengths and valve sizes to be more directly equated to one another. Its units will be given in normal gpm/%. (The N gpm term appeared already in the denominator of the $G_1 G_E$ product in Example 6.2.) For a linear valve, gain is the product of reagent normality times the full capacity of the valve in gallons per minute. (A nonlinear valve would have the same gain but modified by the slope of its characteristic curve as described below.)

Valves are available in several characteristics, but the two most important are linear and equal-percentage. A linear valve, as its name implies, has a constant gain equal to its full flow divided by

100%, through its throttling range, if the pressure drop across it is constant. Due to mechanical limitations, its throttling characteristic deteriorates below a certain flow rate. If adjusted below this point, the valve tends to close completely; if the controller output then is increased, the valve may pop open to its lowest controllable flow. If automatic control is attempted below the throttling range of a valve, a limit cycle will usually result. Some linear valves[4] may be throttled to less than 1% of their capacity—a rangeability exceeding 100:1.

The pressure drop available across a control valve will vary inversely with flow due to the fixed resistance offered by pipes and fittings and the internal resistance of centrifugal pumps. Where reagents can be supplied from head tanks with short lines of ample diameter, little pressure loss will result except at full flow. The effect of pressure loss due to fixed resistance tends to distort the linear characteristic as flow increases, resulting in reduced gain at high flow rates.

An equal-percentage valve will produce a particular *percent* change in delivered flow for a given *increment* in stem position. For example, a change in stem position from 40 to 41% of full scale might change the delivered flow from 1.00 to 1.04 gpm; and a change in stem position from 99 to 100% of full scale would change the delivered flow from 10.0 to 10.4 gpm. Figure 6.6 shows the characteristic of an equal-percentage valve with a 50:1 rangeability. Its equation is

$$f = R^{m-1} \tag{6.18}$$

where f is the fraction full flow, m is the fraction of controller output, and R is the valve rangeability. Differentiation of (6.18) gives the slope of the characteristic curve:

$$\frac{df}{dm} = f \ln R \tag{6.19}$$

The effect of the equal-percentage (logarithmic) characteristic is that valve gain varies directly with flow. If this valve is used without compensation, the loop gain will vary with reagent flow—an undesirable situation.

The one instance where an equal-percentage characteristic is particularly desirable for pH control is the neutralization of a single

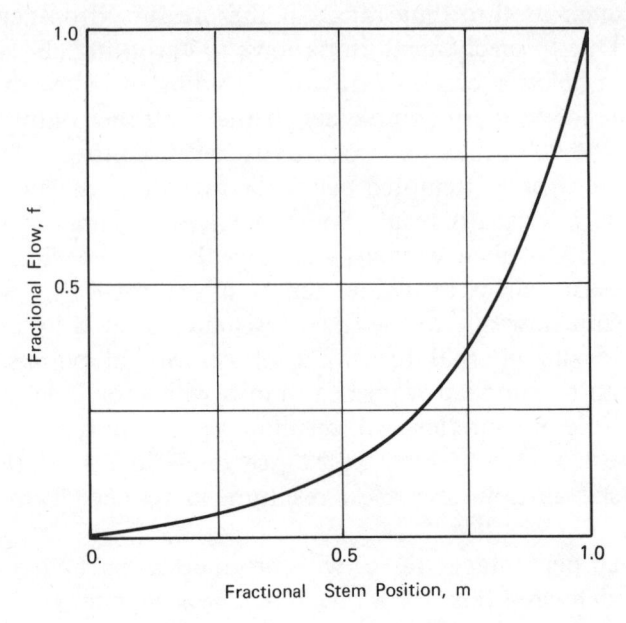

FIGURE 6.6. The characteristic of an equal-percentage valve with 50:1 rangeability.

weak acid or base or its salt. Equations (3.21) and (3.22) derived the slope of a titration curve at a given pH for a weak acid and for a weak base. The slope was related inversely to the concentration of the weak agent in the solution. If that were the only agent to be neutralized, then the flow of neutralizing reagent would be proportional to the concentration of weak agent (at a given flow). In this situation, the direct variation of valve gain with reagent flow would offset the inverse variation of the electrode gain with concentration of the weak agent. The presence of variable amounts of strong acids or bases will tend to distort this relationship.

Some valve styles, such as ball and butterfly, have inherent equal-percentage characteristics. Where the style is desired but the characteristic is not, compensation must be applied, several forms of which are presented in Chapter 8.

Example 6.3

Let a linear valve with a capacity of 0.05 gpm be used to deliver 10% caustic to the process described in the previous examples. The normality of

10% caustic is 2.75 from Table 3.1.

$$G_V = (2.75\ N)(0.05\ \text{gpm}) = 0.1375\ N\ \text{gpm}/100\%$$

$$G_1 G_E G_V = (1120\ \text{pH}/N\ \text{gpm})(0.1375\ N\ \text{gpm}/100\%) = 1.54\ \text{pH}/\%$$

If the range of the pH transmitter is 2 to 12, then the transmitter gain G_T is

$$G_T = 100\%/10\ \text{pH} = 10\%/\text{pH}$$

The gain product for all but the controller is then

$$G_1 G_E G_V G_T = (1.54\ \text{pH}/\%)(10\%/\text{pH}) = 15.4$$

For a loop gain of 0.5, the controller gain would be

$$G_c = \frac{0.5}{15.4} = 0.0325$$

CONTROLLERS

The controller is the one element in the loop which may be readily adjusted to provide stable operation. But because most titration curves are extremely sensitive to reagent addition, the controller gain allowable at the natural period of the loop is usually quite low. To achieve more effective control than would be available with simple proportional action, other control modes are added.

Reset Action

"Reset"[5] is the name applied to integration of the error to produce a change in the manipulated variable:

$$m = \frac{1}{R} \int e\, dt \qquad (6.20)$$

where R is the adjustable reset time constant. Differentiating (6.20) gives the reason why reset action is valuable:

$$\frac{dm}{dt} = \frac{e}{R} \qquad (6.21)$$

The manipulated variable will continue to change as long as an error exists, therefore permanent offset is avoided.

146 Composition Control

Along with this advantage, however, reset adds phase lag to the controller, which increases the period of oscillation of the loop. And as expressed in (6.12), the dynamic gain of the process varies directly with its period, indicating the destabilizing effect of reset.

Applying a sine wave as input to the integrator described in (6.20) results in a cosine wave as an output, whose amplitude varies with its period. In essence, the integrator produces a phase shift ϕ_R and gain G_R of

$$\phi_R = -90°, \qquad G_R = \frac{\tau_o}{2\pi R} \qquad (6.22)$$

Equation (6.22) illustrates again how reset can eliminate a steady-state error—its gain is infinite at an infinite period (steady state).

Derivative Action

Derivative action is achieved by differentiating the error with respect to time:

$$m = D\frac{de}{dt} \qquad (6.23)$$

where D is the adjustable derivative time constant.

The principal role of derivative action is to counter the phase lag of reset with an equal phase lead. Its phase characteristic ϕ_D is therefore of most importance, although derivative gain G_D is also variable:

$$\phi_D = +90°, \qquad G_D = \frac{2\pi D}{\tau_o} \qquad (6.24)$$

Actually, it is impractical to manufacture a differentiator that will reach $+90°$ in phase or produce a gain beyond 15 or 20. However, when combined with the proportional and reset modes, this limitation does not significantly degrade control below what is possible using the ideal formula above.

Combining the Control Modes

Reset and derivative are not normally used alone for control, but are combined with proportional action in a three-mode controller.

The equation for an ideal three-mode controller is

$$m = \frac{100}{P}\left(e + \frac{1}{R}\int e\,dt + D\frac{de}{dt}\right) \quad (6.25)$$

Here P is the proportional band of the controller expressed in percent—it is the change in measurement required to change the output full scale with proportional action alone.

When the modes are combined, their individual contributions must be added vectorially, since they are not in phase with each other. Figure 6.7 gives the vectors representing the gain and phase of each component of output with respect to the error. Since the proportional band modifies all three vectors to the same extent, it is not included in the diagram.

Summing these vectors and applying the proportional gain yields the phase and gain of the controller with respect to period:

$$\phi_c = \tan^{-1}\left(\frac{2\pi D}{\tau_o} - \frac{\tau_o}{2\pi R}\right) \quad (6.26)$$

$$G_c = \frac{100}{P}\sqrt{1 + (\tan\phi_c)^2} \quad (6.27)$$

Studies[6] have indicated that most effective control is achieved when all three modes have unit vectors. This results in zero controller phase and minimum gain at the natural period of the process. To achieve this balance, reset and derivative are set at $\tau_o/2\pi$.

For most processes these settings give the greatest effectiveness of any combination. There is a certain danger in using them for nonlinear processes as described here, however, particularly where certain elements display variable gain. It seems that the loop phase of a single-capacity plus dead-time process with a three-mode controller set as above is in the vicinity of $-180°$ for all periods from τ_o to $10\tau_o$. The loop will not normally oscillate at any period above τ_o, however, because the loop gain is too high, due both to G_1 and G_c increasing with τ_o. But if some other element such as G_E were to decrease substantially, the loop gain could change enough to oscillate at some period longer than τ_o.

148 Composition Control

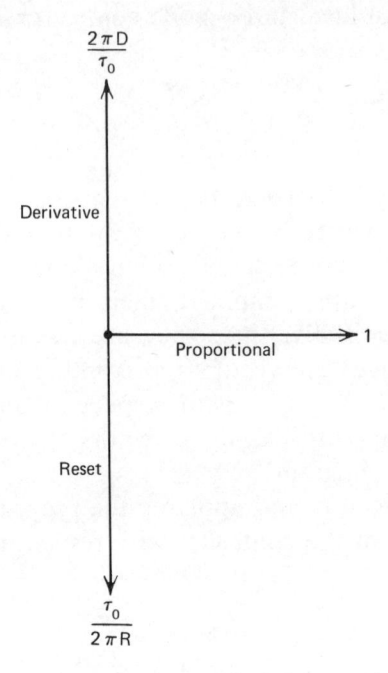

FIGURE 6.7. The vector diagram for an ideal three-mode controller.

To adjust the control modes properly for a neutralization process, determine the natural period of the process by forcing a low-amplitude cycle under proportional control. This is accomplished by setting reset time to maximum and derivative time to minimum, while adjusting the proportional band to promote the desired oscillation. To obtain zero controller phase at τ_o,

$$\frac{2\pi D}{\tau_o} = \frac{\tau_o}{2\pi R}$$

$$DR = \left(\frac{\tau_o}{\pi}\right)^2$$

For stability in the presence of gain variations, R should be set at $4D$. Then

$$D = \frac{\tau_o}{4\pi}, \qquad R = \frac{\tau_o}{\pi} \qquad (6.28)$$

Controller settings may also be estimated from an open-loop measurement of dead time. With the controller in manual operation and the controlled variable at rest, step or pulse the controller output, observing the time required for the first indication of a response. Since τ_o is approximately $4\tau_d$,

$$D = \frac{4\tau_d}{4\pi} = \frac{\tau_d}{\pi}, \qquad R = \frac{4\tau_d}{\pi} \qquad (6.29)$$

In cases where the derivative mode is absent, or cannot be used due to its sensitivity to a high noise level in the controlled variable, reset must be set differently. Here, reset phase lag cannot be avoided, so that the period with reset action is longer than without. Studies[6] have indicated that a controller phase of $-30°$ is optimum in this case, thereby increasing τ_o by 50%. This amount of phase lag is developed when

$$R = \frac{\tau_o}{2.4} = 1.67\tau_d \qquad (6.30)$$

Finally, the proportional band should be adjusted for the desired degree of damping.

Example 6.4

Estimate the settings for a three-mode controller for the process described in the previous examples.

$$D = \frac{\tau_d}{\pi} = 6.4 \text{ sec}$$

$$R = 4D = 25.6 \text{ sec}$$

With the controller adjusted to exhibit no phase lag at the natural period of the loop, its gain is simply $100/P$. Then, from Example 6.3,

$$\frac{100}{P} = G_c = 0.0325$$

$$P = \frac{100}{0.0325} = 3080\%$$

In processes dominated by dead time, derivative is of little value

150 Composition Control

and is not generally used. These processes are more tolerant of reset, however, and a phase lag of 60° has been found optimum.[7] This amount of phase lag also inc-eases the period of the loop by 50% (in the absence of capacity) and is developed when

$$R = 0.14\tau_o = 0.28\tau_d \tag{6.31}$$

A special control mode called "sampled data" has been found extremely useful with dead-time processes. It is described in some detail in Chapter 8.

If the maximum proportional band available is less than that required for damping at the highest electrode gain, a limit cycle will develop whose amplitude lowers the loop gain to unity.

Example 6.5

Given a proportional band limit of 300%, estimate the resulting limit-cycle amplitude:

$$G_c = \frac{100}{300} = 0.33$$

$$G_E = (G_c G_T G_V G_1)^{-1}$$

$$G_E = \left[(0.33)(10)\frac{0.1375}{100}(5.17 \times 10^{-4}) \right]^{-1}$$

$$G_E = 4.22 \times 10^5 \text{ pH}/N$$

From Fig. 6.5, the gain above coincides with pH 5.4 and 8.6. The loop should therefore limit cycle between these two pH values.

Nonlinear compensation can help to equalize the electrode gain for various values of measurement and eliminate the proportional band limit at the same time. This compensation is described in detail in Chapter 8.

Processes with a variable reaction rate will also produce a variable τ_o. Whenever lime slurry or solid is manipulated for pH control this characteristic is likely to appear, since its reaction rate varies with pH, flow, particle size, and carbonate content. To avoid instability in these loops, derivative time should be adjusted to

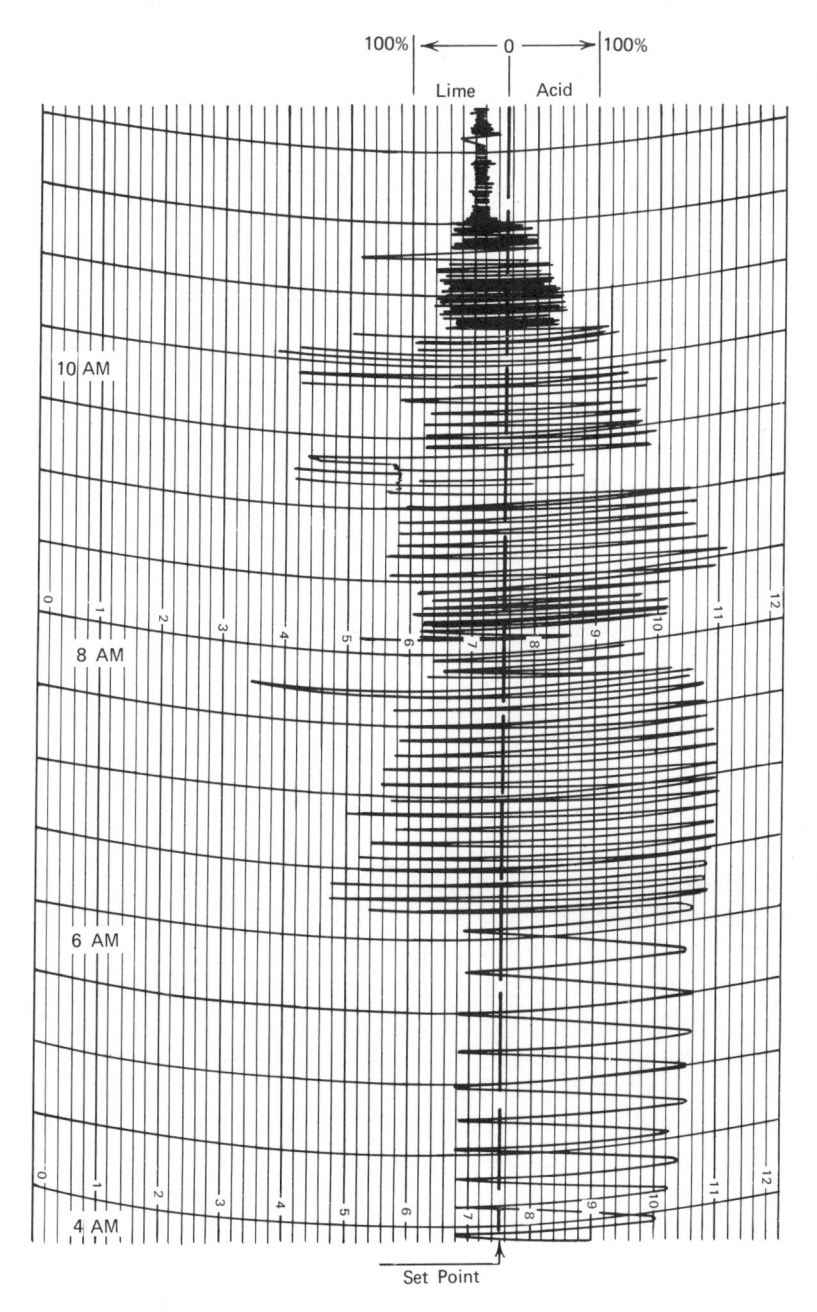

FIGURE 6.8. This is an actual record of pH from a process using a proportional controller manipulating linearized lime and sulfuric acid valves about a set point of 7.5.

about one-fourth of the shortest observed period, and reset to about one-fourth of the longest observed period. Variations in period of 5:1 are not uncommon when using lime as a reagent. Similar characteristics appear with other time-dependent reactions such as the chlorination of cyanide (see Fig. 5.4).

To demonstrate that the adjustment of a pH controller when manipulating lime slurry is not an easy task, Fig. 6.8 is presented as a reproduction of a pH record from such a process. Lime slurry and sulfuric acid were being added through linearized valves manipulated by a proportional controller. This eliminated variable valve gain and controller phase shift from the loop, but variable reaction rate and buffering remained. Although the actual dead time in the process was in the vicinity of 30 sec, the period of oscillation varied from 2.5 to nearly 20 min. With proportional control, valve position is related to deviation from set point, as indicated by the index marked at the top of the chart. As in many waste-neutralization facilities, the composition of the waste at any point in time is unknown and randomly varying, and electrode coating can increase the period with time, such that diagnosis of records like Fig. 6.8 is difficult at best. More is said about controller adjustment under adaptive control in Chapter 9.

SUMMARY

Control effectiveness is maximized when the controller gain is high and the natural period of the loop is short. Both of these factors are ultimately determined by the phase and gain characteristics of the process. Actually, dead time is doubly responsible for a poorly controlled process, since it determines the natural period of oscillation, and through it the dynamic gain of the capacity. Since the controller should be adjusted for stable operation, its mode settings are related to the process period and gain. Ultimately, then, the process itself determines how well it can be controlled, which is the subject of the next chapter.

Some features of ion processes cannot be improved by thoughtful design—notably their extremely nonlinear measurements. Since control effectiveness also hinges on electrode gain, due attention to its compensation is given in Chapter 8.

REFERENCES

1. F G. Shinskey, *Process-Control Systems*, McGraw-Hill Book Company, New York, 1967, pp. 6–17.
2. F. G. Shinskey, *ibid.*, pp. 20–23.
3. T. E. Corrigan and E. F. Young, General Considerations in Reactor Design—II, *Chemical Engineering*, October 1955.
4. C. B. Shuder, Control Valve Rangeability and the Use of Valve Positioners, Instrument Society of America Paper No. 71-817.
5. F. G. Shinskey, *op. cit.*, pp. 12–16.
6. F. G. Shinskey, *op. cit.*, pp. 101, 102.
7. F. G. Shinskey, *op. cit.*, pp. 17, 94.

7

DESIGNING A CONTROLLABLE PLANT

A well-designed control system is only half of what is required for successful regulation of ionic species. If the plant is poorly laid out, without sufficient mixing, or with an unresponsive reagent-delivery system, satisfactory control will be impossible even with the most advanced instrumentation. On the other hand, a plant that is designed to be easily controllable may produce effluent well within specifications with a relatively simple system.

Unfortunately, once a plant is built, modifications to vessels, mixers, and piping are difficult and expensive, if not prohibitive. So it behooves the designer to consider dynamic responsiveness as important as the choice of reagents or materials of construction. Since knowledge of dynamic response is not at present a prerequisite for plant designers or consultants in the waste-treatment industry, most present-day plants are not as well controlled as they could be. But a controllable plant costs no more to build than one that is not, and in fact could cost considerably less, due to attendant reduction in the size of vessels used for neutralization. With this end in view, the various components of the process are examined herein.

MIXING

Mixing is one of the least understood operations in a plant, from the standpoint of dynamic response. Although much has been written on mixer design to retain solids in suspension, generate emulsions, or provide intimate contact between liquids and gases, these are not the problems faced in the control of ion concentration. In general, we are concerned with blending a small amount of relatively concentrated liquid or slurry with a large volume of waste or process water, and detecting its concentration as quickly as possible. The problem is then twofold: blending and transportation to the detector.

Blending

It is absolutely essential that the reagent and the stream to be treated are thoroughly blended—otherwise the resulting concentration will be randomly and uncontrollably variable. Stratification of a dense reagent easily occurs if it is simply poured on the surface of a large stream moving in laminar flow. A representative sample of the mixture cannot be obtained either for feedback control or record of treatment.

Laminar flow at the point of reagent introduction must be avoided. The only alternative would be to distribute the reagent uniformly over the entire cross section of the stream—an impossible task when reagent flow must be varied over orders of magnitude. Turbulent flow consumes energy in the form of mixer horsepower or head loss across a restriction, but it is this energy which accomplishes the blending. The strata of the stream must be sheared and broken, with the reagent filling the voids, in order to produce a homogeneous stream.

In an open channel, turbulence can be caused by baffles and weirs, each introducing a head loss. In a pipeline, the same effect can be achieved with an orifice plate. In each case, the reagent should be introduced immediately upstream of the restriction or outfall, to take advantage of the full head loss. Lateral distribution across a wide spillway obviously must be provided, although unnecessary in a pipeline where turbulent flow exists.

Commercial in-line mixers are available wherein the stream is split, rotated, and recombined several times to produce a uniform blend. These units are designed to operate primarily in laminar flow, absorbing a pressure drop several times as great as pipe of the same diameter. Where sufficient head is unavailable for these passive mixers, motor-driven agitators are offered by some manufacturers for in-line service.

Blending in a conduit minimizes the composition gradient across its width, but the gradient across its length may still vary. Baffles and other in-line mixers have little effect on the longitudinal gradient, so that reagent is, in effect, conveyed by the stream. This process is dominated by dead time, much like a conveyor transporting solids from a gate to the point of measurement. Since the dynamic gain of dead time is unity, control of concentration in a conduit is extremely difficult unless reagent requirements are very low or the solution is well buffered. In most installations, then, capacity must be introduced for the low dynamic gain it affords.

Backmixing

The longitudinal gradient can only be moderated by "backmixing," that is, physically recycling material from output back to input. Backmixing in a pipe is obviously difficult due to frictional losses accompanying the increased velocity at the pipe walls. Therefore, vessels with generally symmetrical dimensions are used to provide averaging, through efficient backmixing of influent with effluent.

The derivation of the dynamic gain of a perfectly mixed vessel was given in the preceding chapter, based on the premise that the concentration was the same everywhere in the vessel. This is exactly opposite to the pipeline dominated by dead time, where a longitudinal gradient existed. However, *perfect* mixing is unattainable, in that *time* is required to transport ions from the point of vessel entry to the point of detection.

For control purposes, backmixing can be envisioned as an equivalent external recycle loop around a vessel as shown in Fig. 7.1. Flow is entering at a constant rate F and leaving the system at the same rate through the action of the level controller. Consider the initial concentration of the discharge to be x_0, at the time the feed

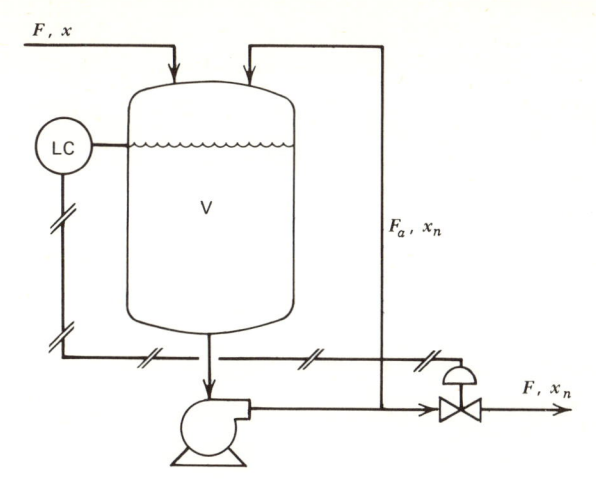

FIGURE 7.1. Backmixing with a pumped recycle loop.

concentration is stepped from x_0 to x_f. If there is no longitudinal mixing within the vessel, that is, plug flow exists, discharge concentration will remain at x_0 for one full displacement of the vessel.

Let the recirculation rate be identified as F_a, with the flow through the vessel the sum $F + F_a$, which is also the capacity of the pump. The transport time through the vessel will then be $V/(F + F_a)$. After one transport time, the concentration in the discharge will step to a new value x_1, as a result of dilution of Fx_f with $F_a x_0$:

$$x_1 = \frac{Fx_f + F_a x_0}{F + F_a} \tag{7.1}$$

Factoring (7.1) yields

$$x_1 = x_f \frac{F}{F + F_a} + x_0 \frac{F_a}{F + F_a} \tag{7.2}$$

To simplify this statement, let α be used to represent $F_a/(F + F_a)$:

$$x_1 = x_f(1 - \alpha) + x_0 \alpha \tag{7.3}$$

This discharge concentration will remain at the x_1 level until a second transport time passes, yielding x_2 as Fx_f diluted with $F_a x_1$:

$$x_2 = x_f(1 - \alpha) + x_1 \alpha \tag{7.4}$$

158   Plant Design

Substituting for x_1 from (7.3) yields

$$x_2 = x_f(1-\alpha) + [x_f(1-\alpha) + x_0\alpha]\alpha$$

Combining terms:

$$x_2 = x_f(1-\alpha^2) + x_0\alpha^2 \qquad (7.5)$$

The procedure may be repeated as each transport time passes, producing the general solution

$$x_n = x_f(1-\alpha^n) + x_0\alpha^n \qquad (7.6)$$

where n is the number of transport times passed since the initiation of the input step. It is related to time t as

$$t = n\left(\frac{V}{F+F_a}\right) = n\frac{V/F}{1-\alpha} \qquad (7.7)$$

Equation (7.6) may be rearranged so that α appears just once:

$$x_n - x_0 = (x_f - x_0)(1-\alpha^n) \qquad (7.8)$$

This, then, is the general step-response equation for recirculation through a vessel in plug flow.

To facilitate comparison between different values of α, x_n of (7.8) is tabulated as x_t, using time expressed in units of vessel residence

Table 7.1 Response of the Backmixed Model $(x_t - x_0)/(x_f - x_0)$

			α		
$\frac{t}{V/F} = n(1-\alpha)$	0.9	0.8	0.7	0.6	0.5
0.1	0.1	—	—	—	—
0.2	0.19	0.20	—	—	—
0.3	0.27	—	0.30	—	—
0.4	0.34	0.36	—	0.40	—
0.5	0.41	—	—	—	0.50
0.6	0.47	0.49	0.51	—	—
0.7	0.52	—	—	—	—
0.8	0.57	0.59	—	0.64	—
0.9	0.61	—	0.66	—	—
1.0	0.65	0.67	—	—	0.75

time V/F. The tabulation is only valid for integer values of n, however. Table 7.1 summarizes the relationship.

The response for the case where $\alpha = 0.8$ is plotted against time in Fig. 7.2. A smooth curve is also shown passing through the average or midpoint of each step. Since longitudinal dispersion will occur in any real vessel, some averaging will ensue, so that the concentration at the discharge will not develop a staircase but a smooth curve. If some particles are retained in the vessel longer than one transport time, others must exit in less time to maintain the average. Consequently the actual dead time in response to a step input can approach half the transport time, as the curve indicates. This property is corroborated by tests in full-scale reaction vessels in which the observed dead time was always less than the vessel volume divided by the pumping rate of the mixer.[1]

Figure 7.3 compares the average response curves for $\alpha = 0.5, 0.8$, and 1.0. The curve for $\alpha = 1.0$ is the familiar exponential step response formulated in (6.5) for the perfectly mixed vessel. Observe that all curves pass through 63% complete response at time equal to the residence time of the vessel, V/F.

The imperfectly mixed vessel can then be represented by dead time equal to half the transport time, and capacity which makes up the balance of the residence time:

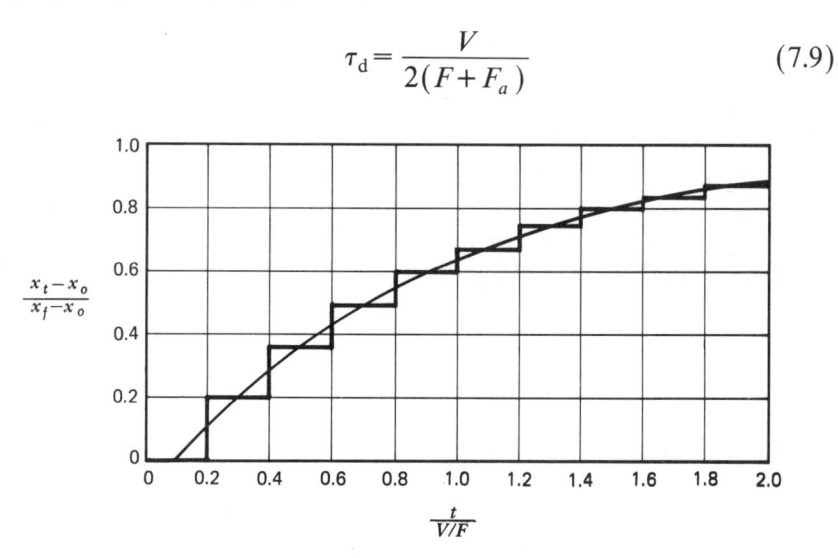

$$\tau_d = \frac{V}{2(F+F_a)} \tag{7.9}$$

FIGURE 7.2. Step response of the backmixed vessel with $\alpha = 0.8$.

160   Plant Design

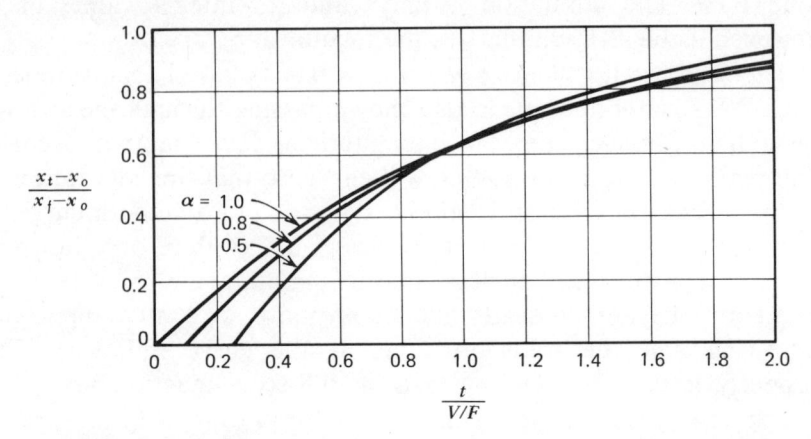

FIGURE 7.3. All curves pass through 63% complete response at $t = V/F$.

$$\tau_1 = \frac{V}{F} - \tau_d \tag{7.10}$$

Again, the average particle must remain in the vessel for one residence time, part of which time it may be in plug flow, and the remainder backmixed.

Choice of Mixers

An external pumped recycle is not customarily used for mixing, due to frictional losses within the pump and recycle line. Instead, an agitator is used within the vessel, but its effect is identical, in principle. Flow patterns are necessarily important, however. A low-speed paddle impeller, for example, can pump a large volume of material radially about the shaft, without effective mixing. A vortex may be formed, in laminar flow. Blending of reagent with the vessel contents thus tends to be poor, and dead spaces exist in corners.

Higher speed axial agitators (300 rpm and up) with marine propellers or pitched turbines are recommended to develop the turbulence necessary for blending. Vertical baffles should be used in a cylindrical tank to avoid vortex formation. A rectangular tank may not need baffles because its corners break the circular flow pattern. Off-center or off-vertical mounting of the agitator also helps. The resulting turbulent flow necessarily increases power requirements, but pockets are virtually eliminated.

A handbook[2] published by The Bethlehem Corporation intended as a guide for the selection of its agitators provides information most consistent with the author's experience. For mixing fluids of low viscosity, that is, water, a speed of 1750 rpm is recommended up to 1000 gal of vessel capacity, and 420 rpm above that volume. Their statement on costs is worth quoting:

> The cost of an *Ergulator* [trademark for Bethlehem agitators] goes up as its speed goes down because of the need for speed-reduction mechanisms, heavier shafts, larger bearings and bigger seals. Accordingly, propeller-type *Ergulators*, where applicable, are cheapest, followed by turbine types, with slow-speed paddles and similar variations becoming more costly.

Horsepower required for "medium agitation" estimated using the nomograph in the handbook for certain size vessels is presented in Table 7.2. Once having determined the required horsepower and speed, a proper size propeller may be selected. Beyond 25 hp, propellers are no longer practical and pitched turbines are used.

Experience in neutralization reactions conducted with medium agitation has been good, yielding τ_d to V/F ratios around 0.05. By contrast, mild agitation yields τ_d to V/F ratios in the 0.1 to 0.2 range. Even with no agitation, the ratio will tend to be in the 0.2 to 0.3 range simply due to diffusion, friction, and changes in direction of flow. The ratios just mentioned apply strictly to vessels whose dimensions are similar to one another—a long vessel such as a trench or pipe will exhibit proportionately more dead time. Although it is obviously desirable to reduce dead time as much as possible, an economic optimum does exist. Dead time can only be reduced at the expense of increased flow, and in a fixed fluid regime, the required power varies as flow cubed—placing a practical lower limit on τ_d.

Table 7.2 Horsepower Requirements for Medium Agitation

Vessel Volume (gal)	Horsepower	Hp/1000 gal
100	0.27	2.7
1000	2.3	2.3
10,000	12	1.2
100,000	80	0.8

162   Plant Design

Compressed air may be used for agitation within certain geometrical limits. Since air leaving a sparger rises to the liquid surface vertically, lateral mixing is poor. Consequently air agitation is restricted principally to vertical vessels of narrow cross section. Some reactors are designed this way, but waste-treatment plants are usually spread out horizontally to minimize head losses and associated pumping costs. Therefore air cannot be recommended for backmixing the contents of these vessels.

Mixing and Dynamic Gain

To permit correlation of the ratio of dead time to residence time with the backmixing factor α, the following derivation is presented:

$$\frac{\tau_d}{V/F} = \frac{V}{2(F+F_a)}\left(\frac{F}{V}\right) = \frac{F}{2(F+F_a)}$$

$$1 - \alpha = 1 - \frac{F_a}{F+F_a} = \frac{F}{F+F_a}$$

$$\frac{\tau_d}{V/F} = \frac{1-\alpha}{2} \tag{7.11}$$

The importance of backmixing may be envisioned by examining (6.11) for the dynamic gain of a first-order lag

$$\lim_{\phi_1 \to -90°} G_1 = \frac{1}{F}\frac{\tau_o}{2\pi\tau_1} \tag{6.11}$$

If the dead time is short relative to the lag, then

$$\lim_{\phi_1 \to -90°} \tau_o = 4\tau_d$$

If we assume that all the dead time in the loop is due to imperfect mixing, (6.9) and (7.10) may be substituted into (6.11):

$$\lim_{\phi_1 \to -90°} G_1 = \frac{1}{F}\frac{4\tau_d}{2\pi(V/F - \tau_d)} = \frac{1}{F}\frac{2}{\pi(V/F\tau_d - 1)}$$

Next, substitute (7.9) into the above:

$$\lim_{\phi_1 \to -90°} G_1 = \frac{1}{F} \frac{2}{\pi\{[V2(F+F_a)/FV]-1\}}$$

$$\lim_{\phi_1 \to -90°} G_1 = \frac{2}{\pi(F+2F_a)} \cong \frac{1}{\pi F_a} \qquad (7.12)$$

As can be seen, the volume of the vessel ultimately has no effect on its dynamic gain, which is primarily determined by the agitator.

Equation (7.11) could also be substituted into the equations above, to give dynamic gain in terms of the backmixing factor, α:

$$\lim_{\phi_1 \to -90°} G_1 = \frac{1}{F} \frac{2}{\pi[2/(1-\alpha)-1]}$$

$$\lim_{\phi_1 \to -90°} G_1 = \frac{2}{\pi F}\left(\frac{1-\alpha}{1+\alpha}\right) \qquad (7.13)$$

Equation (7.13) is only meaningful as α approaches 1, to satisfy the limit of ϕ_1. (The gain is not $1/F$ with $\alpha = 0$ as it would be for a pure dead-time process.) Therefore further simplification is justified if the limit is placed on α:

$$\lim_{\alpha \to 1} G_1 = \frac{1-\alpha}{\pi F} \qquad (7.14)$$

This expression is useful as a "rule of thumb" for rapid estimates of loop gain. For a nominally 90%-backmixed vessel ($\alpha = 0.9$) the dynamic gain ratio is approximately 0.033 and for a 95%-backmixed vessel, half that value. The following example will justify its use.

Example 7.1

Determine α for the process described under Example 6.1 and estimate the dynamic gain using (7.14):

$$\frac{V}{F} = \frac{200 \text{ gal}}{100 \text{ gpm}} = 120 \text{ sec}, \quad \tau_d = 10 \text{ sec}.$$

$$\frac{\tau_d}{V/F} = \frac{10 \text{ sec}}{120 \text{ sec}} = 0.0833$$

164 Plant Design

Using (7.11),

$$\frac{1-\alpha}{2} = 0.0833$$

$$1-\alpha = 0.1667, \qquad \alpha = 0.833$$

$$G_1 \approx \frac{0.1667}{3.14(100 \text{ gpm})} = 5.3 \times 10^{-4}/\text{gpm}$$

Actually, the gain calculated in Example 6.1 is inexact, in that V/F was chosen for τ_1 instead of $V/F - \tau_d$. If the latter had been used, a slightly higher gain would have resulted:

$$G_1 = \left(\frac{1}{100 \text{ gpm}}\right)\left(\frac{39}{2\pi 110}\right) = 5.64 \times 10^{-4}/\text{gpm}$$

VESSEL SELECTION

Agitation is not the only important aspect of plant design, although it plays a major role in vessel design. Long, narrow vessels are to be avoided if effective backmixing is to be achieved, with cylindrical or cubic vessels being more desirable. Locations of entry and exit are also important, but too frequently the selection of vessel size itself is not given adequate consideration.

Residence Time

The discussion on backmixing concluded that the dynamic gain of the process is unrelated to volume, which left the selection of residence time open. However, this conclusion was based on the assumption that all the dead time was due to imperfect mixing. The designer is guarded against going to the extreme of selecting a residence time approaching zero because then the other elements in the loop become important. If a sample must be withdrawn from the reaction vessel and sent to an externally mounted electrode assembly, dead time will be added to the loop bearing no relationship to the vessel residence time. And valve response, although usually in the order of 1 sec when small amounts of reagent must be delivered, is also a factor. There may additionally be some dead time in delivering reagent from the valve, although proper piping design can minimize this. The electrodes themselves do not respond instantaneously, and this could be the limiting factor in some situations. Reference 3 reports time constants from 0.25 to 8 sec, varying with buffering and the direction of pH change. Coatings, however, can increase this to minutes.

One time limitation stands out above all others, however, and this must be the prime consideration in the design of any vessel where a reaction will be conducted. If insufficient time is allowed for the reaction to go to completion, the measured concentration will not represent the final state of the system and control will not be effective.

Where soluble reagents such as caustic or hydrochloric acid are added to neutralize influent streams whose active components are also soluble, the rate of reaction seems to be virtually instantaneous. In these cases, the vessel's residence time can be reduced to a very low value with acceptable results. The extreme "well-mixed vessel" is a centrifugal pump with reagent added to the suction and pH measured at the discharge as shown in Fig. 7.4. Due to high pressure and velocity at the pump discharge, a sample is usually withdrawn, rather than have the electrodes inserted directly in the pipe. This adds dead time to the loop, but response can be very rapid, with a natural period of only a few seconds. Then any downstream capacity can easily attenuate disturbances and low-

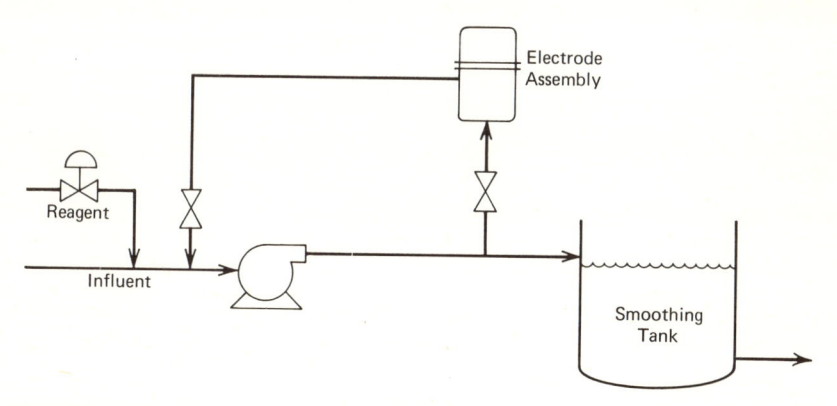

FIGURE 7.4. The well-mixed vessel with a minimum residence time is the centrifugal pump.

amplitude oscillations. This technique has been used successfully in cases where reagent demand is moderate or the stream is well buffered. It is certainly worth investigating in applications where a centrifugal pump is already part of the process.

Reaction Rate

Some reactions are not instantaneous, however, as in the reduction of chromium(VI) (see Fig. 5.5). Neutralization reactions with an insoluble reagent such as lime also require time for completion, as dictated by the solubility of the particles. The pH electrode will respond to the hydrogen ions present, but not to the particles of undissolved reagent. If insufficient residence time is allowed, the pH of the effluent can still be controlled at 7, for example, but could later reach as high as 10 or 11, given enough time for complete reaction.

Simply reducing the set point to a lower value will not substitute for insufficient residence time. If, in the same example, the pH is controlled at 5, the effluent may eventually reach pH 7 while the influent enters at pH 3. If the influent were to enter at pH 5, however, no lime would be required for control, and the final state of the effluent would also be pH 5.

The rate of a chemical reaction depends on the concentration of the unreacted component or components. In a "first-order" reaction this dependency is limited to one component, either because it is the only reactant, as in a decomposition, or because its concentration in the continuous phase is limited compared to other species present. In any event, first-order reactions are quite common, and more complex reactions can often be represented as first order through at least part of their concentration range.

The rate of reaction[4] is simply described as a rate of change of concentration of the controlling reactant, as a function of its concentration:

$$\frac{-dx}{dt} = kx \qquad (7.15)$$

Here, k is the first-order rate constant expressed in units of inverse time. If the reaction is conducted batchwise, concentration of the controlling reactant will vary exponentially with time as indicated by the integration of (7.15):

$$\int_{x_0}^{x} -\frac{dx}{x} = k\int_0^t dt$$

$$\ln x_0 - \ln x = kt \qquad (7.16)$$

Subscript zero refers to conditions at the start of the reaction.

In a second-order system, the rate of consumption of either reactant A or B varies with the product of their concentrations:

$$\frac{-dx_A}{dt} = kx_A x_B \qquad (7.17)$$

Integration similar to that performed for the first-order system can be used to obtain the concentration versus time relationship. Clearly if one reactant is in appreciable excess, its concentration will not significantly change with time and the equation reduces to first order, with the other reactant controlling.

For continuous reaction in a backmixed vessel, the dynamic material-balance equation applies. The sum of the reactant fed, plus its change in inventory in the vessel, equals the rate of withdrawal:

$$Fx_0 + V\frac{dx}{dt} = Fx \qquad (7.18)$$

Here x_0 represents the concentration of reactant in the feed and x is its concentration within and leaving the reactor. Substituting for dx/dt from the first-order reaction (7.15),

$$F(x_0 - x) = Vkx$$

$$x = \frac{x_0}{1 + kV/F} \qquad (7.19)$$

Given a fixed residence time and rate coefficient, x varies directly with x_0. To state it a little differently, an increased rate of reaction can only be achieved by increasing the amount of unconverted reactant in the effluent.

Figure 7.5 plots the time response of hydrochloric acid neutralized batchwise with lime slurry.[5] The curves do not conform to any

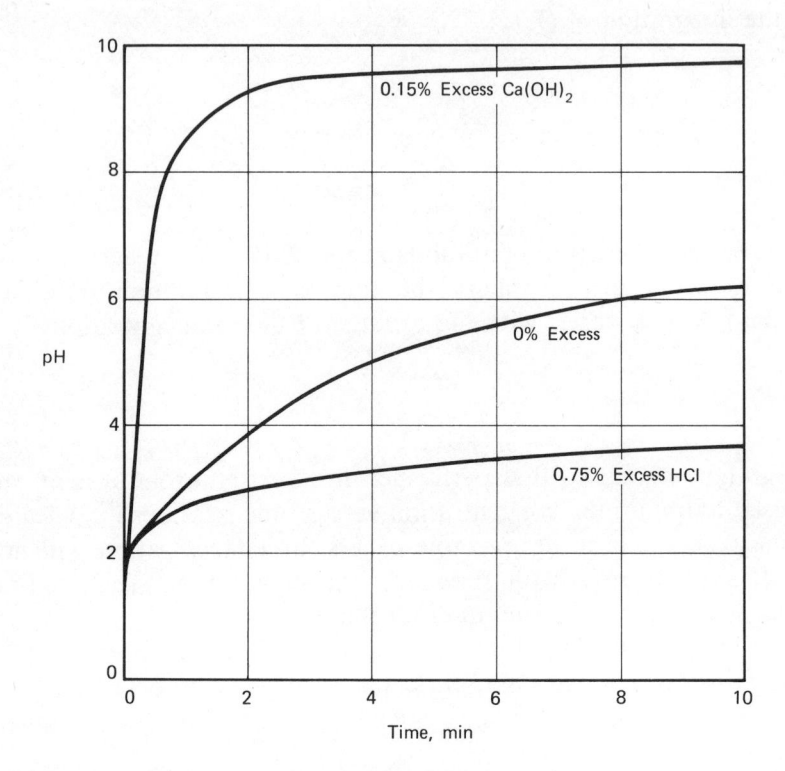

FIGURE 7.5. The reaction rate between hydrochloric acid and lime slurry is strongly dependent on a slight excess of lime.

of the conventional reaction rate equations, the principal difference being the marked dependency of rate upon excess lime. An explanation for this phenomenon could be that all lime slurries contain perhaps 1% $CaCO_3$, due either to incomplete calcining or absorption of CO_2 from the air. (Absorption does take place readily, as observed by rapidly developing cloudiness in freshly prepared limewater exposed to the air.) A final pH of 9 or 10 would prevent the carbonate from dissolving at all, while at pH values below 8, it will slowly continue to react with the H^+ ions present.

The slower reaction rate below pH 8 could be avoided by conducting a first-stage reaction at pH 9.5 as shown in Fig. 7.6. This product could then flow into a second stage where acid influent could be added to reduce the pH to the desired neutral level.

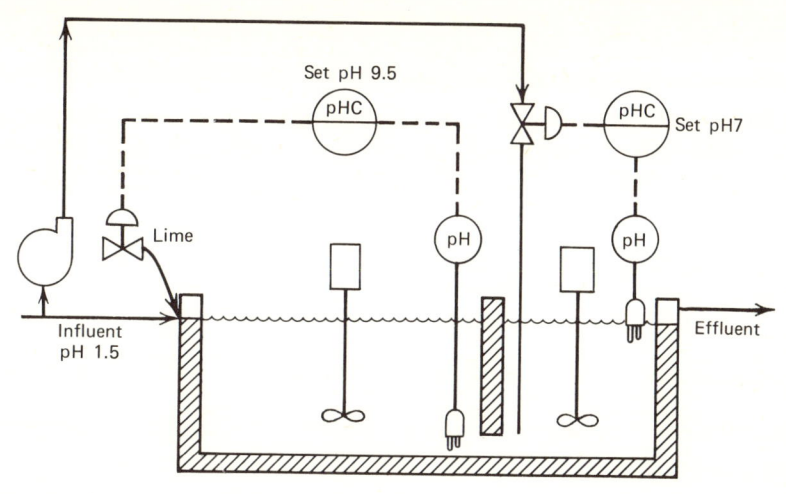

FIGURE 7.6. Neutralization in two stages can take advantage of reaction rate increasing with lime concentration.

Operating the second stage in this way has two distinct advantages:

1. Effluent pH will respond rapidly to acid flow.
2. The difficulty in finding a suitable valve for manipulating a small flow of lime slurry is avoided.

Single-state neutralization with lime in a residence time of 15 min has been satisfactorily carried out with influent pH of 3 and above. For influents requiring 100 times more lime, or where reduction in residence time is important, the two-stage system in Fig. 7.6 is recommended.

Vessel Arrangement

Influent and reagent should enter at the same point, premixed for best results. The direction of inflow should oppose the direction of agitation for most effective backmixing. Since axial flow agitators must pump downward (to avoid spraying the vessel contents out the top) the optimum point of entry would be at the bottom center. While this may be provided with small elevated tanks, most waste-treatment vessels are installed in ground, of poured concrete, making that point inaccessible.

Hoyle[6] has found that dead time is lower if the influent and reagent enter at the surface of an in-ground tank rather than at the bottom near a wall. The observation is made that the agitator circulates flow downward around its shaft and upward near the walls, where flow would enter. Then flow introduced at the bottom of a wall would be swept upward along the wall first, having to negotiate a longer mean path than flow entering at the surface, as shown in Fig. 7.7.

The exit should be diametrically opposite to the point of entry, to minimize short-circuiting. Otherwise a significant part of the influent could pass by the vessel without treatment, and uncontrollable variations in effluent quality could result. If the influent enters at the bottom on one side, effluent should leave at the top from the other side, and vice versa, as shown in Fig. 7.7.

Two-stage in-ground vessels must have alternate overflow and underflows since the first-stage effluent is the second-stage feed. In Fig. 7.6, flow enters the top of the first stage and leaves the top of the second.

In neutralization of industrial wastes, many streams with widely differing properties are usually combined for treatment in a single vessel. Often both acid and basic reagents are required when the influent pH may be on either side of 7. In these instances, reagent can be saved by providing a retention vessel upstream of the neutralization tank. This smoothing vessel should have enough residence time to accommodate the largest expected transient influent variations. A trade-off exists between the capital investment

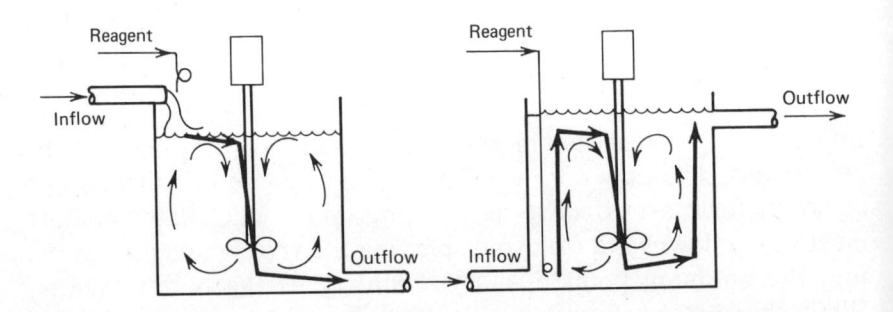

FIGURE 7.7. Influent and reagent entering at the top have a shorter mean path to the exit, due to the downward circulation of the agitator.

for the tank and savings in reagent which it affords. Each situation would have to be evaluated on its own merits.

Even when a single reagent is used, a smoothing vessel can be helpful. Occasionally, heavy loads may be placed on the system for short intervals of time, when ion-exchange beds are backflushed, when spills occur, or during periodic rinsing of vessels. To accommodate these severe upsets may require reagent-delivery capability far in excess of normal demands. A smoothing vessel can help shave these peaks, using a reasonably sized reagent system.

Superior response will be achieved if the pH control loop has a short period of oscillation, and downstream capacity is used to attenuate transients. So a smoothing vessel is also useful downstream, as shown in Fig. 7.4. An application of this technique is given in Reference 1. Agitation does not need to be provided in smoothing vessels, because they are not inside a control loop. Dead time may account for 20 to 30% of the residence time, but the first-order lag still dominates, providing almost as much attenuation as would be obtained with vigorous mixing.

Properly locating the point of measurement is as important as locating the vessel exit, and is, in fact, tied to it. A representative measurement of effluent quality is essential and so the electrodes should be placed in the effluent stream. However, they should not be placed out of the mixed zone. Dynamic response is also important, so every effort should be made to maintain a reasonable flow past the electrodes. Avoid the use of stilling wells or other protective devices which restrict flow. Also avoid locations where flow is variable.

Submersible electrode assemblies will always be more responsive than flow-through assemblies since they are directly inserted within the process stream. Flow-through assemblies were designed for pressurized service with a sample withdrawn continuously for measurement. Although desirable for easy maintenance, the additional delay caused by the sample line impedes control action.

Submersible electrodes are used primarily in open vessels. They should extend only slightly below the surface of the liquid to minimize the possibility of leakage. Occasionally, however, deeper submergence is necessary, as shown in Fig. 7.6. The assemblies may be air-purged to protect against leakage even under severe conditions, as described in Fig. 2.10.

172   Plant Design

Flow-through electrode assemblies are required for pressurized service, or whenever a sample must be treated prior to measurement. Sample lines should be as short as practical and velocity high to minimize dead time. High velocity also helps keep the electrodes clean, although excessive velocity can cause erosion and even cleavage of the electrodes. A velocity of 2 ft/sec past the electrodes is probably as high as tolerable for long life.

Another problem appears when an organic-phase sample must be analyzed. The sample is first extracted with water, followed by decanting of the organic phase. If both organic and aqueous phases are allowed to contact the electrodes, their charge difference prevents a steady reading from being obtained. Figure 7.8 describes a simple decanter for such an application. The organic overflow must be at a slightly higher elevation than the aqueous to achieve the proper separation. An interface will form in the vertical leg, such that its level below the two overflows is in proportion to the ratio of the two densities, ρ_1 and ρ_2:

$$\rho_1 h_1 = \rho_2 h_2 \qquad (7.20)$$

Dimension $(h_2 - h_1)$ must be selected to keep the interface above the tee.

Protection against Failure

Even when great care is exercised in designing a waste-treatment plant and its controls, in selecting and adjusting components, and in maintaining the equipment, accidents will still happen. An electrode

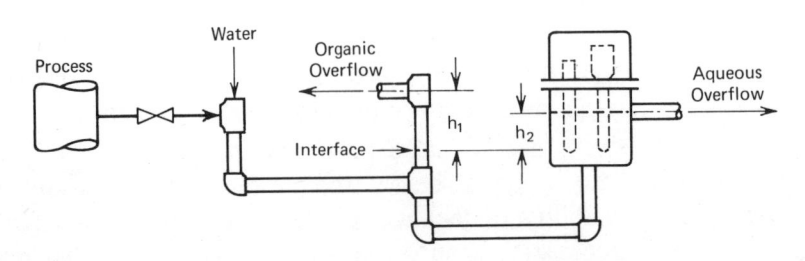

FIGURE 7.8. An organic-phase overflow must be provided when extracting with water.

could suddenly fail, an air compressor could trip out, a valve could plug, a reagent tank run dry, or an agitator or pump lose its impeller. Or even with all equipment in perfect condition, the plant could sustain a temporary overload during which time control over the effluent is lost.

In some plants, even a temporary loss of control can result in damage to the environment. If the effluent is discharging directly into a biological treating system, the bacteria population could be destroyed; or if discharging directly to a stream, fish and other aquatic animals and plants may be killed. Even if the effluent is discharged to a sewage system, the escape of untreated or overtreated material could release toxic gases such as hydrogen sulfide or cyanide into pumping stations or treatment plants. With proper plant design, most of these accidents are avoidable.

The first requirement of any protective system is a place to put off-specification waste. There must be a vessel somewhere in the system that is normally empty or at least has enough spare capacity to hold up to 2 hr accumulation of effluent. This time interval ought to be long enough to restore control or repair the defective electrode or whatever caused the failure. Or if repair cannot be accomplished, it ought to be sufficient to begin an orderly shutdown of those

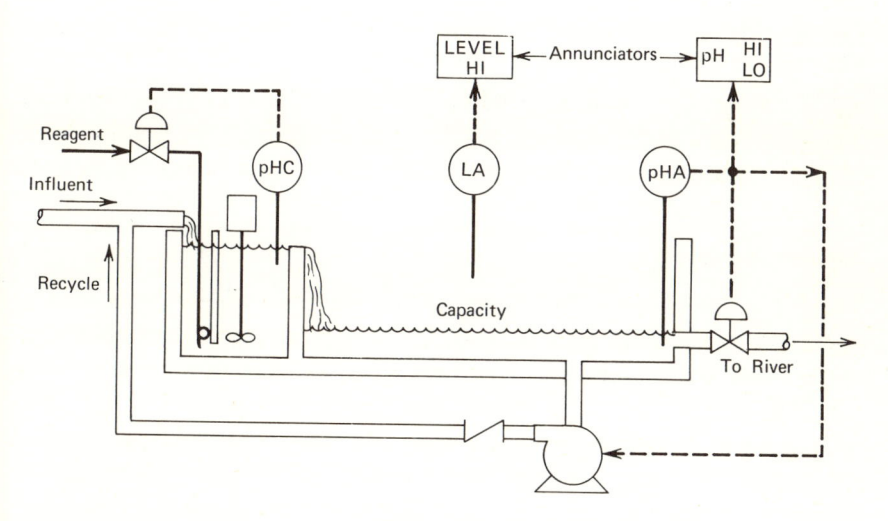

FIGURE 7.9. Emergency capacity must be provided with a means to recycle off-specification waste.

process units which are the heaviest contributors to the effluent system. Sizing of this storage capacity should be based on an intelligent evaluation of procedures necessary to shut down the plant in the event of such an emergency.

Having a place to store off-specification waste is only half the solution to the problem, however. Provision should also be made for recycling it for retreatment. Figure 7.9 shows a storage vessel with spare capacity located in the path of the treated effluent. Failure of the effluent pH to meet the prescribed high or low limit triggers an alarm which closes the discharge valve. Unless reagent is added to this vessel, there is no way that off-limit material can be corrected. At the same time as the discharge valve is closed, then, a recycle pump may start to return the waste for retreatment. This will not lower the level in the storage vessel, however, since overflow from the reaction tank will increase due to recycling. A high-level alarm will have to warn the operator of impending overflow so that corrective action may be taken to curtail inflow.

REAGENT DELIVERY

Reagent has presumably been selected on the basis of economic and safety requirements. For example, lime is usually chosen over caustic for neutralization because of its lower cost, but also because it is easy for operators to handle without fear of burns. Unfortunately, however, it presents delivery problems that caustic does not. The problems and requirements for proper delivery of soluble and insoluble reagents are presented in the paragraphs below.

Soluble Reagents

These are the simplest to deliver accurately over wide ranges, but a few basic rules must be followed. First, the line downstream of the valve must not be allowed to drain freely. Otherwise, reagent will continue to flow into the vessel after the valve has been shut, and will be slow to start again after the valve has been opened. Figure 7.9 shows the most reliable method of avoiding line drainage—a loop seal in the line near its exit. Then the control valve may be mounted in a remote location without sacrificing dynamic response. Check valves have also been used for this purpose but may eventually leak in severe service. A small-diameter tube should be used to allow the line to be purged of air after installation. If above the surface, the loop seal prevents air from backing into the line; if submerged, it prevents the less dense effluent from displacing the more dense reagent.

For atmospheric systems, gravity flow of reagent from a head tank is satisfactory. For pressurized operation, a pump is required which can cause problems of its own. Because rangeability requirements are usually quite high, the pump must be sized above the average flow. Throttling or recycling of the excess capacity can lead to considerable heat release, which for concentrated acids can cause severe corrosion. Pressurized storage under an inert gas head is a preferred method of supplying reagent. Metering pumps with adjustable stroke can also be used for pressurized systems, although their characteristic is linear and rangeability is limited. Valves permit more flexible operation with extended rangeability. In any system wherein soluble reagents are used, filters should be provided to protect the small internal parts of valves from damage or plugging.

Insoluble Reagents

Hydrated lime [$Ca(OH)_2$] is the most popular of the insoluble reagents and will be used as an example throughout this discussion. Actually it is moderately soluble, the saturated solution having a pH of about 12.53 at 25°C. The solution is obviously too weak to be used to neutralize strong acids, and so the lime is fed either dry or as a slurry. Where small quantities are required, it may be procured in 100-lb bags. It may also be obtained in 35% slurry form in tanktrucks or cars as a by-product of acetylene manufacture. The properties of both are quite similar in their reaction with acids.

In tonnage quantities, quicklime (CaO) may be shipped by truck in 1/2-in. size for slaking on the site. Water is added to hydrate the lime in the slaker, and heat is released as steam, with a thick slurry removed from the slaker. Rocks and sand are to be expected when using this source of lime. Velocities in slurry lines have to be high enough to avoid settling of the lime, yet low enough to minimize erosion of valves and fittings by the sand.

All the limes cited above are called "high-calcium," to distinguish them from "dolomitic" limes which contain equal parts of magnesium, for example $CaO \cdot MgO$ and $Ca(OH)_2 \cdot MgO$. The latter are not as reactive[7] as the high-calcium limes since magnesium hydroxide is a weaker base than calcium hydroxide and is even less soluble.

Limestone ($CaCO_3$) may also be used for neutralization, but reacts very slowly. Its use is largely reserved for emergency treatment of acid spills in beds of broken limestone or clam shells. Recently, however, limestone is gaining more favor for neutralizing acid wastes and also for SO_2 removal from flue gases, both in the dry and wet states. To be effective, however, it must be ground to a very fine particle size (e.g., -200 mesh). In large-scale installations, ball mills are being included with cyclone classifiers to grind and deliver the required fine slurry.

Particle size is only one limitation to the reaction, however—the very insolubility of limestone promises a slow rate even with consistently fine particles. As a consequence, some unreacted limestone will always leave with the effluent, the amount depending on its pH and residence time. Although this will mean an increase in reagent consumption (possibly significant), it can also mean loss of control if the concentration of acid in the influent is too high. Butler[8] gives a pH of 9.8 for $CaCO_3$ saturated in distilled water. But the pH for saturation decreases with the acidity of the solution dissolving the $CaCO_3$ due to the accumulation of Ca^{2+} ions. For $BaCO_3$ (with essentially the same solubility as $CaCO_3$), Butler gives a pH of 6 for a saturated solution in 0.01 N acid, increasing about one pH for every decade reduction in acid normality. In other words, a strong acid at pH 2 can only be neutralized to pH 6, or pH 3 to 7, and so on, with $BaCO_3$. If the controller's set point were higher than this saturation pH, it would open the valve wide, adding more reagent without affecting the pH. Excess reagent could dissolve later, however, in water containing less ions, and raise the pH of that water substantially. Consequently limestone cannot be recommended for complete neutralization of wastes whose acid content exceeds about 0.001 N.

Dry lime may be added directly to the surface of the vessel where a slurry is being prepared or even where the neutralization is taking place. The latter is obviously desirable in that it avoids slurry-handling problems, although it has some limitations. Dry lime does not wet readily, and so tends to linger on the surface for several seconds after addition, unless forced below by a spray of liquid. This can add dead time to the control loop, and so spraying the influent on the surface should be considered when directly neutralizing with dry lime.

The lime feeder is a modulated hopper from which an auger removes lime at a controlled rate. The mechanism may be driven by a constant-speed motor operating on a percentage of a time cycle, or by a variable-speed drive. In either case, rangeability is limited to about 20:1, although the maximum rate is manually adjustable.

When the top of the vessel is not open for dry-lime feeding, the feeder is used to drop lime into an integral slurry tank, to which a proportionate amount of water is added. One of these systems is shown schematically in Fig. 7.10.

In many pH control systems, water is not proportioned to the

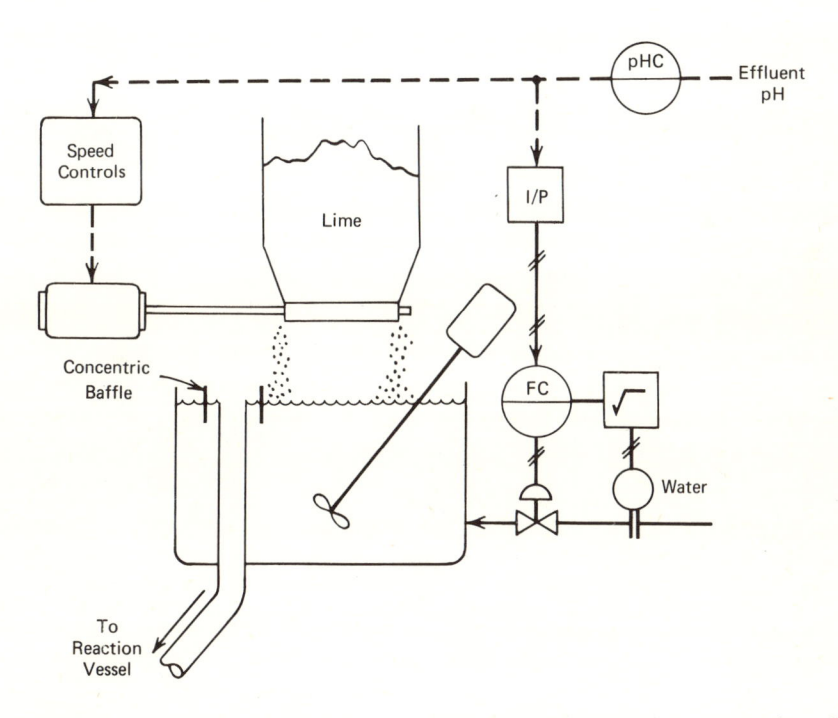

FIGURE 7.10. Proportioning of lime and water is essential to responsive pH control.

178   Plant Design

lime but is added at a constant rate. This is detrimental to effective control, because the slurry tank then becomes a large time constant in the worst possible place—between the manipulated variable and the reaction vessel. Consider what would happen in the case of a sudden reduction in load. The effluent-pH controller would cut back the lime addition rate sharply, but lime slurry would continue to be added until the concentration in the slurry tank was exhausted. With the system shown in Fig. 7.10, however, water flow is adjusted by the pH controller, changing the rate of overflow directly, to bring about responsive control.

In cases where gravity overflow of slurry is not possible, a pumped recirculation loop must be provided, for flow cannot be stopped without danger of plugging. Piping runs should be smooth and carefully arranged to avoid trouble spots. Connections for flushing should be provided and all taps should be on vertical runs or on the top of horizontal runs. Control valves should be located at the high point in their individual lines to avoid plugging when closed. Pressure gauges mounted on the top of a 6-in. nipple of 1-in.

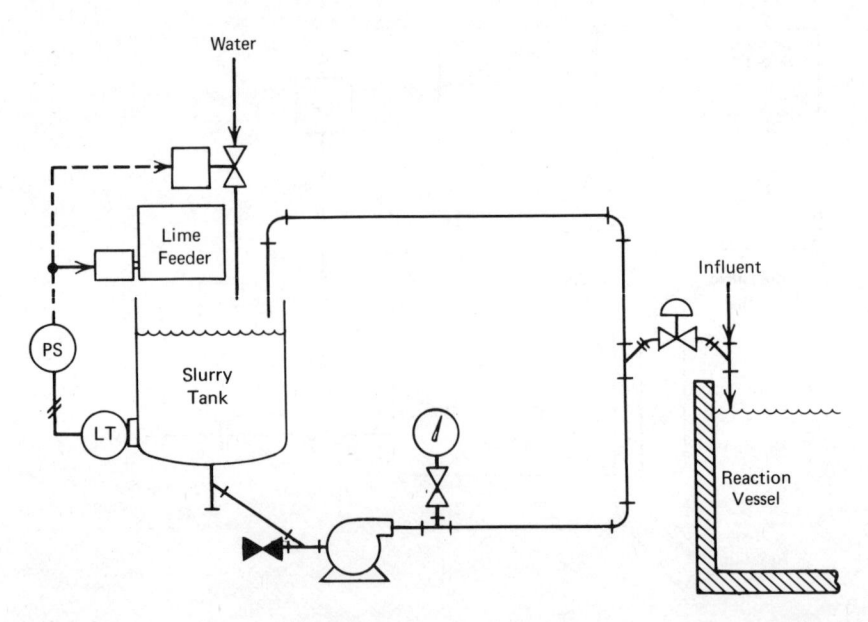

FIGURE 7.11. Careful piping design is essential when manipulating the flow of a slurry.

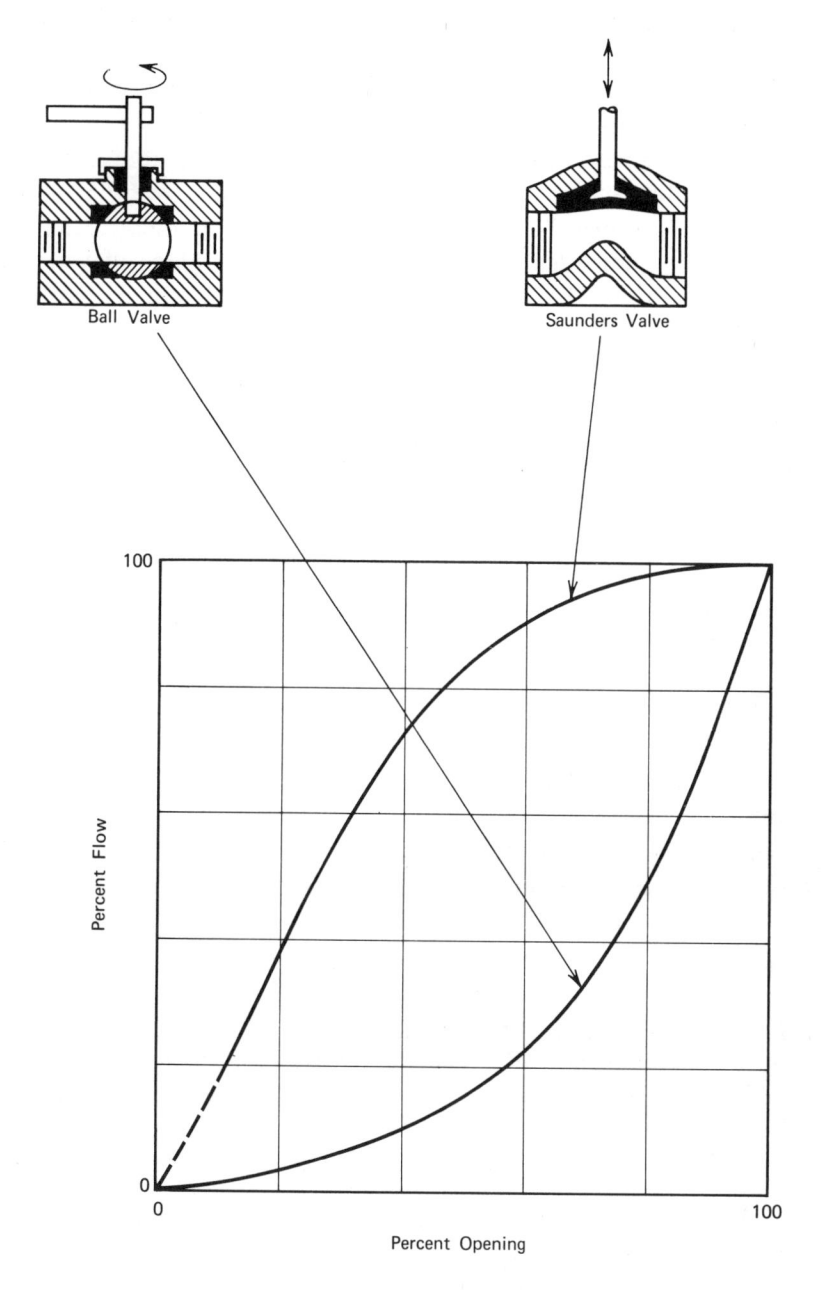

FIGURE 7.12. The ball valve has an essentially equal-percentage characteristic, while that of the Saunders valve is on the opposite side of linear.

179

180 Plant Design

pipe will not be susceptible to plugging. All of these features are described in Fig. 7.11.

Primary control over reagent addition in this case is with the slurry control valve. Then slurry preparation can be performed batchwise, using a constant-speed feeder and a solenoid valve for water. A switch acting on the level signal can energize both on low level, and deenergize both on high level. In any of these examples, treated effluent or even influent may be used in place of water to prepare the slurry. This is especially attractive where the influent pH is very low (0–2), since the maximum slurry concentration that can be easily circulated is about 12% solids.

Ball valves are probably most satisfactory for throttling lime slurry. They are equal-percentage by nature, so external characterization is necessary if linear response is desired. At this writing, ball valves are not available for control below 1/2 in. in size. Saunders-type valves, wherein a flexible diaphragm is positioned over a curved weir, are also used for throttling slurries. They are quite dependable, but their rangeability is 10:1 or less and their characteristic resembles that of a quick-opening valve, not desirable for pH control. These two valves and their characteristics are shown in Fig. 7.12.

Metering pumps have been used on lime slurry with varying degrees of success. Suspended particles tend to interfere with the action of check valves in suction and discharge, with resultant slippage. Also low velocities tend to promote settling, so careful piping design is still required.

REFERENCES

1. F. G. Shinskey and T. J. Myron, Adaptive Feedback Applied to Feedforward pH Control, Instrument Society of America Paper No. 565-70.
2. Bethlehem Ergulator Selection Handbook, The Bethlehem Corporation, Bethlehem, Pennsylvania.
3. A. L. Guisti and J. O. Hougen, Dynamic of pH Electrodes, *Control Engineering*, April 1961.
4. T. E. Corrigan and E. F. Young, General Considerations in Reactor Design—II, *Chemical Engineering*, October 1955.
5. P. J. Docherty, Jr., Automatic pH Control: Neutralization of Acid Wastes by Addition of Lime Slurry, Masters Thesis, Dartmouth College, Hanover, New Hampshire, June 1972.

6. D. L. Hoyle, The Effect of Process Design on pH and pIon Control, presented at the 18th National Symposium of the Analytical Instrumentation Division of The Instrument Society of America, San Francisco, May 3–5, 1972.
7. W. A. Parsons, *Chemical Treatment of Sewage and Industrial Wastes,* National Lime Association, Washington, D.C., 1965.
8. J. N. Butler, *Ionic Equilibrium: A Mathematical Approach,* Addison-Wesley, Reading, Massachusetts, 1964, pp. 255–258.

8

FEEDBACK CONTROL SYSTEMS

The previous two chapters developed the theory of feedback control and the characteristics of real plant equipment. In this chapter we will "put it all together" with instrument components especially selected to match the process. This "putting together" is recognized today as *systems engineering*—the design of control systems from available components to achieve some optimum plant performance.

First, however, the objectives of the plant must be recognized before components can be selected which will meet them. Then, after having designed a system, an evaluation must be made of its anticipated performance, to determine whether it will meet the stated objectives. If it will not, more powerful control techniques such as feedforward must be called upon, or the plant redesigned to make it more controllable.

THE CONTROL PROBLEM

Whether the stream whose composition is to be controlled is a salable product or a waste effluent, certain specifications must be met under threat of penalty. In a production operation, failure to control composition precisely may mean low yield, poor recovery, excess recycle, or other debits. But penalties for out-of-specification plant wastes can be even more severe, including destruction to life or governmental litigation. In either case, the objective of the control system is complex—a certain degree of control must be achieved over a broad range of operating conditions. If plants were always in a steady state, automatic control would be unnecessary, so it is the range of loads encountered and their rate of change which tax the control system. The first order of business is then to examine the sensitivity of the controlled variable to disturbances in reagent and influent flow rates.

Accuracy

Measurement of composition using potentiometric methods exhibits a variable sensitivity due to the logarithmic nature of the electrode potential. In regions of high concentrations where the sensitivity is low, a better means of measuring composition should be sought. For most applications, however, low concentrations of excess reactants are to be measured and controlled, and here the sensitivity of the method is unsurpassed. By the same token, aquatic life is just as sensitive to the presence of the measured ions, making close control in this highly sensitive range mandatory.

The most sensitive titration curve is that between a strong acid and a strong base. To control at pH 7 ± 0.5 requires that acid and base concentrations be within $\pm 0.28 \times 10^{-6}$ N of one another. How accurately reagent must be added depends on how much must be added. If the initial pH is 6 or 8, concentration must be adjusted by $0.99 \times 10^{-6} \pm 0.28 \times 10^{-6}$. Manipulation of reagent flow with an accuracy of $\pm 28\%$ will meet the objective. If the initial pH is 4 or 10, however, concentration must be adjusted by $10^{-4} \pm 0.28 \times 10^{-6}$ or $\pm 0.28\%$. Making a small adjustment to the pH of process water may therefore be accomplished with ease, whereas neutralizing a more concentrated waste stream may be extremely difficult to control.

In an effort to place the problem in perspective, consider the limits pH 7 ± 0.5 to be a target 6 in. in diameter. Bringing a waste stream from pH 4 to 7 would compare to hitting the target from a distance of 100 ft. A starting point of pH 2 would be about 2 miles away, and pH 0 would be equivalent to a distance 200 miles from the target.

Loop gain may be seen here as a distinct function of valve size. Neutralizing a pH-4 waste would require a valve 100 times as large as needed to adjust pH-6 process water, yet with the same absolute accuracy.

The presence of weak acids or bases, or buffers such as carbonates, can reduce the loop gain and accuracy requirements by orders of magnitude. But by the same token, the process may be less predictable. So a controller adjusted for a lightly buffered condition may not be sufficiently responsive when heavier buffering is encountered.

Rangeability

Rangeability refers to the variation in flow of reagent which must be controllably delivered. Reactions conducted as a part of manufacturing a product ordinarily do not require a wide range of flow rates. They are operated at some optimum residence time and profitable production rate, using feedstocks of relatively consistent quality.

Treating of wastes, however, is another matter. Usually there is a single waste-treatment facility for an entire plant, handling a mixture of many streams containing a variety of chemicals. If the plant manufactures piecework, or if some of its areas operate discontinuously, wide-ranging upsets are likely to be encountered. For strong acids and bases, the reagent demand can be calculated from pH measurements. Most waste-treatment facilities must cope with mixtures of strong and weak agents, however, including various metal ions. In these cases, reagent requirements can best be determined by evaluating a representative set of titration curves.

The principal problem posed by wide-ranging reagent demand is the limitations of feeding equipment. Metering pumps and dry feeders are limited to about 20:1 turndown, while conventional control valves will only throttle over a range of 50:1. If forced to throttle below their minimum rated flow, all these devices tend to shut off entirely. This action will cause a sharp change in the controlled variable, sending the controller output back into the throttling range, where flow may be too high for the current demand.

A limit cycle is thereby produced. The period of the cycle will be the natural period of the control loop unless the reagent-delivery system exhibits hysteresis at its low limit. A metering pump which stops feeding when its control signal falls to 5% and does not resume until it increases to 10% will produce a much slower cycle than one without this property. The phase shift produced by hysteresis varies with the amplitude of the control signal[1]:

$$\phi_H = -\sin^{-1}\frac{H}{A} \qquad (8.1)$$

where H is the hysteresis bandwidth and A is the peak-to-peak amplitude of the input signal.

The amplitude of the limit cycle appearing in the controlled variable can be estimated knowing that the manipulated reagent flow is cycling continuously between zero and its minimum controllable flow:

$$A_c = A_m G_1 G_E \qquad (8.2)$$

Here A_c and A_m are the amplitudes of the controlled variable and reagent flow, and G_1 and G_E are the process dynamic gain and electrode gain as defined in Chapter 6. If the period of the loop increases due to the phase contribution of hysteresis, G_1 will increase proportionately. The limit cycle will not necessarily be sinusoidal since reagent flow is manipulated stepwise. In the presence of hysteresis, where controlled throttling occurs through part of the cycle, the controlled variable generally traces a sawtooth waveform.

Not infrequently, control is required over four decades of influent pH, for example, pH 2 to 6 or 8 to 12. In absence of buffering, this means delivering reagent over a range of 10,000:1. This extreme rangeability requirement is peculiar to waste-treatment facilities and is explored in more detail later in this chapter.

A critical review of the influent characteristics under all phases of plant operation is needed to predict reliably the rangeability demanded of the reagent system. Simply because a waste stream has a pH as low as 2 does not mean that it may be as high as 4 or 5 for any length of time. For example, during periods of normal operation the pH may vary only from 2 to 3 or thereabouts; but if various areas of the plant are shut down overnight or weekends, discharges from the equipment remaining in service could be in the range of pH 4 to 5. If this is the case, the control system would require a rangeability covering pH 2 to 5.

When the influent varies on either side of neutrality, its pH could conceivably be anywhere between two extremes. However, the likelihood of a nearly neutral influent may be remote. Consider an influent comprised of an acid waste at pH 2, and a caustic waste of pH 12, mixed in varying amounts. The probability of the mixture being between pH 4 and 10 is low—the two streams would have to be balanced within 1 part in 100! Only during shutdown of *both* major sources of waste would more neutral solutions be encountered.

Nonlinearity

One example of the extreme nonlinearity which may be encountered in ion control was given in the titration curve of Fig. 6.4. If control is exercised about neutrality, the nonlinearity is symmetrical. However, if the control point is at pH 4 or 10, symmetry is lost and effective compensation is more difficult to achieve. The most difficult requirement would be the need to change the control point over several pH units, in an effort to adjust the rate of an equilibrium reaction.

Weak agents also affect the symmetry of the titration curve. A 1.0 N solution of acetic acid exactly neutralized with caustic will have a pH of 9.35, which is the midpoint of its titration curve. The midpoint shifts with influent concentration, however, being pH 8.85 for 0.1 N acetic acid and pH 7.0 for 0.0 N acid. Metal ions such as chromic, ferrous, and zinc act as weak acids with a sharp rise in their titration curve from pH 3 to 5 and another from pH 7 to 9. The nature of the titration curve in the process in question therefore should be evaluated before attempting to apply nonlinear compensation.

When small adjustments in pH are made, the nonlinearity appears much less severe. Figure 8.1 is the same titration curve as that traced in Fig. 6.4, except reduced in range by three decades at each end. Observe that the reagent scale has been reduced 1000-fold

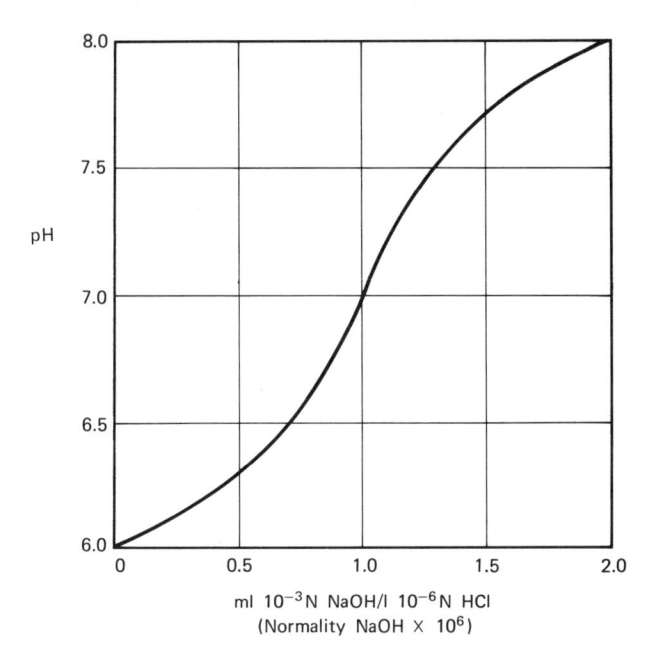

FIGURE 8.1. A narrow operating range moderates the nonlinearity of a titration curve.

as well. The gain change between pH 6 and 7 is only about 2:1. This is in complete agreement with the gain curve given in Fig. 6.5, indicating that the same gain curve is applicable over any influent range, although only for a single control point.

SPECIAL COMPONENTS

The three outstanding characteristics of ion-controlled processes just cited call for techniques not required in more conventional control loops. By the same token, new components had to be developed or adapted to meet these challenges. A valve with a rangeability of 10,000:1 could perhaps be developed, but at high cost, with but a limited market in view. In this case conventional valves and other selected components were combined in such a way as to achieve the objective. Fortunately, the extreme demands of these strange processes can be satisfied with a modest array of readily available components.

188 Feedback Control Systems

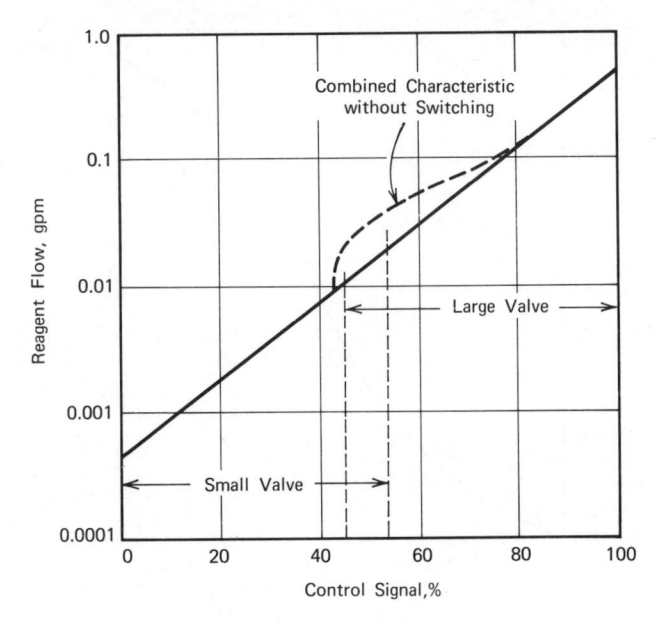

FIGURE 8.2. The two equal-percentage valves must be coordinated to act as one.

Achieving Wide Rangeability

Logically, two valves, each with 50:1 rangeability, would seem capable of being sequenced to achieve nearly 2500:1 overall rangeability. This is easier said than done, however, bearing in mind that the size of the smaller valve would have to be one-fiftieth that of the larger. If the valves were of linear characteristic, sequencing would be awkward, in that the controller-output signal would only change from 0 to 2% to stroke the smaller valve fully.

Equal-percentage valves can be readily sequenced with almost equally balanced operating ranges, however. Refer to Fig. 8.2 to see how this is done. On semilogarithmic coordinates, the maximum flow of the larger valve is marked at the 100% output line. Next the minimum flow from the smaller valve (its maximum flow divided by its rangeability) is marked on the 0% output axis. Then the two points are connected with a straight line. Locating the maximum flow of the smaller valve and the minimum flow of the larger valve will indicate the signal ranges over which these valves must operate.

Their respective valve positioners must be calibrated to these exact ranges.

The valve positioner is a proportional controller which compares the control signal against the valve stem position and adjusts the motor to minimize their difference. It may be calibrated to drive the valve full stroke over any span of control signal beyond about 10%, even with reversed direction, if desired. It also overcomes the 3 to 5% hysteresis normally present in a pneumatic valve motor.

Note the overlap between the two valves indicated in Fig. 8.2. Valve sizes are available in certain irregular increments only, so that the two valves could not be sized in exactly a 50:1 ratio. But even if two sizes in a 50:1 ratio were available, some overlap is always desirable to reduce the frequency of transferring control between the two valves. As a result, the overall rangeability for a sequenced pair must always be less than the product of their individual rangeabilities.

If the smaller valve is allowed to remain open when the larger valve is opened, the reagent flow will be double the desired amount. As the larger valve is opened more, the flow from the smaller valve will continue to provide a portion of the total, even after it has reached its limit. The result is a distorted relationship between flow and the control signal, indicated by the broken curve in Fig. 8.2. To avoid this discontinuity, only one valve must be allowed open at a time.

Figure 8.3 shows a pneumatic sequencing system wherein a pressure switch selects the operating valve based on the value of the controller output. The switch must be set to energize the solenoids when the smaller valve has reached full stroke, and deenergize them when the larger valve has fallen to its minimum controllable flow. Its "lockup" or "differential gap" is then equal to the overlap between the valves. In practice, the valves may take a second to open and close, but the process is not fast enough to respond to such a short transient. Yet if switching is not provided, the on-off action of the large valve when adjusted below its throttling range will cause a limit cycle at that load level.

Figure 8.3 also shows a pneumatic 1:1 amplifier located between the pressure switch and the solenoid valves. It is necessary to avoid a momentary reduction in the air signal while the solenoid valves are changing position, as this will cause the switch to chatter.

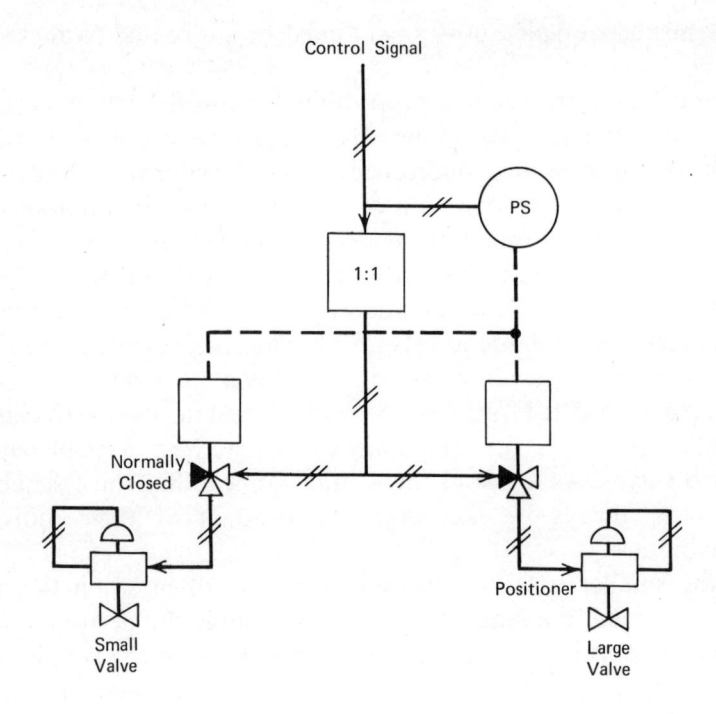

FIGURE 8.3. The pressure switch will allow only one valve to be open at a given time.

Infinite rangeability can be achieved in systems delivering both acid and basic reagents. Consider each reagent delivered by a metering pump whose output can be adjusted controllably down to 5% of full flow. If their flow is limited at that point rather than shut off completely, reagent delivery essentially becomes the difference between the two flow rates. Neither pump would then be required to operate below its controllable limit, therefore allowing infinite rangeability. Consuming both reagents at the same time is the price paid.

A ball valve[2] has been introduced on the market with a claimed rangeability of 1000:1. The ball is slightly notched at its edge so that the orifice presented to the fluid at low flow rates is not the usual lens opening. This extends the control range over a few more degrees of rotation.

The Trim Controller

For rangeability demands beyond 1000:1, a third valve is necessary if limit cycling is to be avoided. In many cases, the limit cycle caused by adjusting the smaller valve below its throttling range may be acceptable, being less than that induced by the large valve by the ratio of their sizes. If a third valve is required, however, it is more easily operated independently by a proportional controller, rather than be sequenced with the other two valves. This controller would typically add base over an influent pH range of 5 to 7 or acid over pH 7 to 9. Two of these trim valves could be operated from a single proportional controller if there were a need for both reagents. In this case both valves would be closed at midscale output from the controller, with the base valve opening on a decreasing signal and the acid valve on an increasing signal.

The valve manipulated by the proportional controller would also have an equal-percentage characteristic. The reason for this choice is that a deviation in effluent pH will cause a change in loop gain (see Fig. 6.4) unless suitably compensated. The equal-percentage valve has a gain which varies with stem position in a logarithmic manner, tending to offset the gain variation of the process. The valve gain is low owing both to its small size and to its characteristic, such that the proportional band of the controller may be reduced to about 10%. The offset incurred should be at an acceptable level without reset or derivative action.

Reset cannot be used in two controllers on the same measurement—otherwise the smaller valve will eventually be driven fully open or closed. Furthermore, if the pH controller has reset action, an equal-percentage valve will not properly compensate for the variable electrode gain, since the position of the valve is no longer proportionally related to pH, but floats with the load. Reset action would return the pH to the set point regardless of what valve position may be required to meet the load. Valve gain then would vary with load.

Valve Characterizers

Since the use of an equal-percentage valve with a controller having reset action results in a variable loop gain, then the equal-

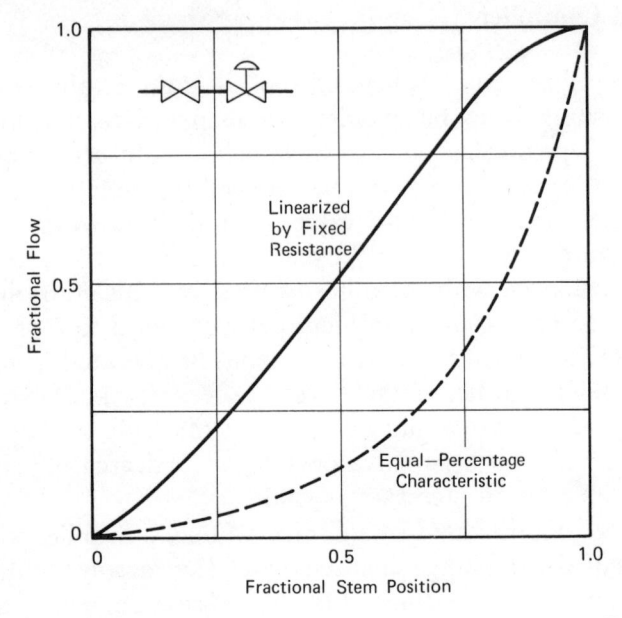

FIGURE 8.4. A fixed resistance in series with the valve can linearize its characteristic.

percentage characteristic must either be avoided or compensated. There are times when it cannot be avoided, as when two valves must be sequenced for wide rangeability. In addition, linear valves are not available in some sizes and styles.

A simple way to compensate for the logarithmic valve characteristic is to insert a fixed resistance in the reagent line. The curve of Fig. 8.4 shows the effect of adding a series restriction whose capacity is one-fourth of that of the wide-open control valve. A fixed orifice may be used, or a hand valve throttled to develop 16 times as much pressure drop as the control valve does at full flow. (The ratio of 16 is used since pressure drop varies as the square of flow.) Obviously the control valve must be sized four times as large as it would be with no series resistance.

The series resistance also reduces the effective rangeability of the valve. Maximum flow is limited by the resistance to one-fourth of the valve's full capacity, whereas minimum flow is not affected at all. Therefore the ratio of maximum to minimum flow (the flow rangeability) is one-fourth that of the valve itself.

Certain styles of positioners can be used to alter a valve's characteristic. Stem position is fed back to the amplifying unit with a cam, which may be contoured to convert a linear valve toward equal-percentage or equal-percentage toward linear. The available cams are not as nonlinear as the equal-percentage characteristic, however.

Compensation can also be provided with an analog divider. This device solves the equation:

$$B = \frac{A}{Z+(1-Z)C}$$

where B is the output and A and C are inputs, all expressed in fractions of full scale. Coefficients Z and $(1-Z)$ are the zero and span adjustments which may be applied to input C. Connecting input A to input C yields a nonlinear function of input A:

$$B = \frac{A}{Z+(1-Z)A} \tag{8.3}$$

The curve thus generated is the opposite of the equal-percentage valve. Differentiation yields the gain of the device for values of input A:

$$\frac{dB}{dA} = \frac{Z}{[Z+(1-Z)A]^2}$$

The gain changes continuously from $1/Z$ at $A=B=0$, to Z at $A=B=1$. The gain change is then $1/Z^2$. An equal-percentage valve with rangeability R will exhibit a gain change of R from minimum to maximum flow. To apply proper compensation using a divider, then, Z is selected according to R:

$$\frac{1}{Z^2} = R$$

$$Z = \frac{1}{\sqrt{R}} \tag{8.4}$$

Solid-state electronic curve characterizers could also be used to generate the nonlinear function, but their gain is usually limited to 5 to 10, as with cams. When manipulating both acid and base valves from a single controller, however, this is the best choice for developing the required "S"-shaped curve.

The Nonlinear Controller

If uniform damping is to be achieved in a highly nonlinear pH control loop, a complementary nonlinear control function must be used. The simplest form of this nonlinear function appears as a combination of three straight lines as shown in Fig. 8.5. Both the width of the deadband and the gain within it must be adjustable to match the particular process being encountered.

As with the process titration curve, the controller gain is not simply the slope of the nonlinear function at a given point, but rather the slope of a line connecting that point with zero error. In mathematical terms, the gain G_f of the nonlinear function $f(e)$ is:

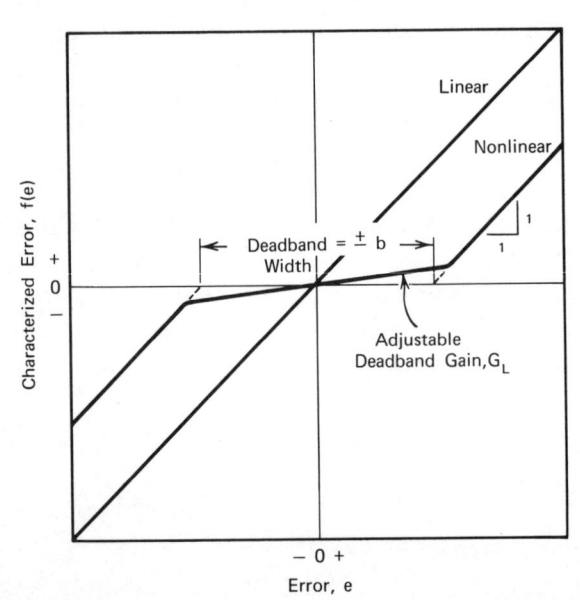

$$G_f = \frac{f(e)}{e} \neq \frac{df(e)}{de} \qquad (8.5)$$

FIGURE 8.5. The deadband width and gain are independently adjustable in a nonlinear controller.

where e and $f(e)$, the error input and its function, are expressed in the same units.

The nonlinear function can be described as

$$f(e) = (e-b) > G_L e$$

where b is the width of half the deadband and G_L is the gain within the deadband. Then the gain G_f is

$$G_f = \frac{|e|-b}{|e|} > G_L \tag{8.6}$$

Example 8.1

To illustrate the improvement in loop gain possible with the nonlinear controller, multiply values of G_f calculated above by G_E given in Fig. 6.5. Use $G_L = 0.02$ and $b = 1.5$ pH units, with a set point of pH 7. While the gain change for G_E is 72:1 over the pH range from 4 to 10, the product of the two gains changes only 3:1 over the same range.

pH	G_f	G_E	$G_f G_E$
7	0.02	2.17×10^6	4.23×10^4
6 or 8	0.02	1.01×10^6	2.02×10^4
5 or 9	0.25	2.0×10^5	5.0×10^4
4 or 10	0.50	3.0×10^4	1.5×10^4
3 or 11	0.63	4.0×10^3	0.25×10^4

The nonlinear function must be adjusted to fit the particular process being controlled. In the absence of a titration curve, or when the buffering is variable, this becomes a trial-and-error procedure. With the deadband set at zero, a limit cycle will ordinarily ensue, even with a very wide proportional band. The width of the limit cycle is a guide—the deadband need not be quite as wide to stop the limit cycle.

An unbuffered system will require the lowest available setting of G_L. If G_L is too low for a given process, however, it can actually promote a limit cycle exceeding the width of the deadband. The following discussion on conditional stability is necessary to demonstrate this point.

196   Feedback Control Systems

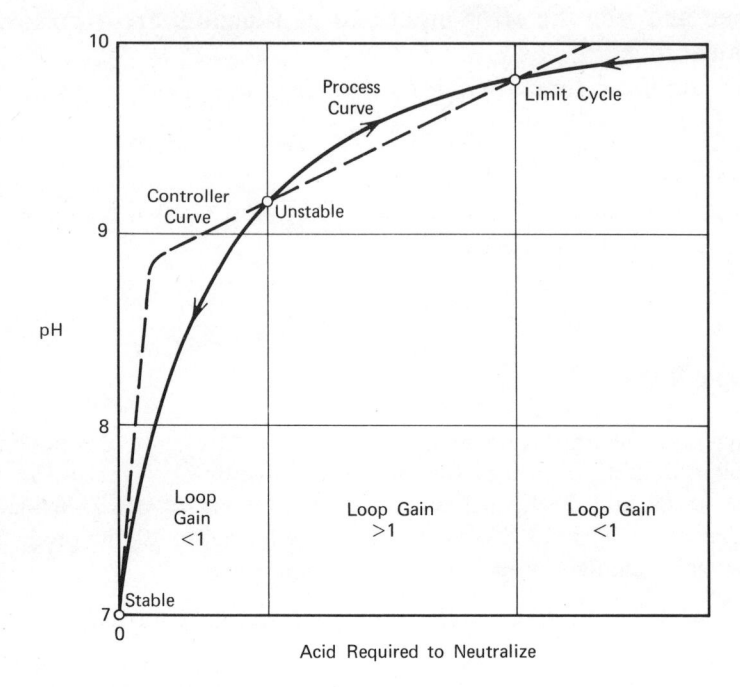

FIGURE 8.6. Conditional stability with a nonlinear controller on a pIon process.

Inasmuch as it is impossible to match a smooth titration curve with the piecewise function of the nonlinear controller, different conditions of stability may exist. Compare the process curve and the controller nonlinear function in Fig. 8.6. (Only one side of the curves is shown—assume the opposite sides identical for ease of explanation.)

In this example, the controller has been adjusted so that a loop gain of 1.0 is crossed twice—at pH 9.2 and 9.8. Should the pH be below 9.2 (and above 4.8) the loop gain is less than 1.0 and a damped return to the set point will follow. But should an upset drive the pH beyond 9.2 (or 4.8), the controller will overcorrect, causing an expanding oscillation. It can only expand to an amplitude of 4.2 to 9.8, however. At that point the loop gain again passes through unity: Larger excursions are attenuated while smaller ones are expanded. The result is a limit cycle with a fixed amplitude of 4.2 to 9.8.

Such a loop could be perfectly stable as long as upsets were small. But once the unstable point is crossed, the ensuing limit cycle will not be broken without intervention. The arrows drawn on the process curve indicate the direction that the oscillation will take in each region.

The discussion above on conditional stability is important because a nonlinear controller can cause a large-amplitude limit cycle if its gain is too *low*! When a linear controller is used on a pH process, it tends to develop a limit cycle at the natural period of the loop, because the loop gain is greater than 1.0 for small deviations from set point. But a nonlinear controller with too low a gain within the deadband will allow the pH to drift out of the deadband until a region of higher gain is encountered. Since this is a drifting condition, it takes longer than the same variation in pH caused by control action. When the controller gain has finally increased enough to turn the pH back toward the set point, the elapsed time tends to be much longer than the natural period of the loop. The pH then drifts in the other direction, past the set point (since the gain is too low to force convergence) into the high-gain region on the other side. The pH traces a limit cycle which exceeds the deadband in amplitude and is much longer than the natural period.

Figure 8.7 compares this limit cycle to the more common one formed by a linear controller on a pH process. With the linear

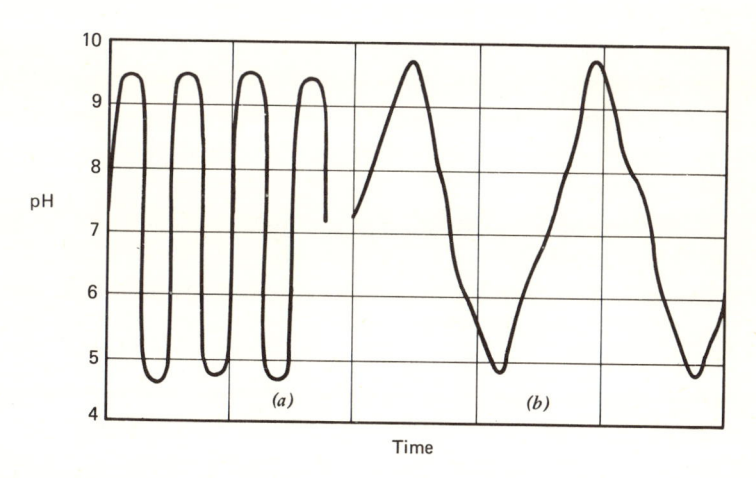

FIGURE 8.7. Limit cycles caused by: (*a*) a linear controller; (*b*) a poorly adjusted nonlinear controller.

controller, loop gain is highest near the set point, so the velocity is greatest there. With the nonlinear controller, the waveform is nearly triangular, since the loop gain is low in the region around the set point and higher farther away. The amplitude of each cycle is primarily a function of controller settings. In the case of the linear controller, amplitude *increases* with a *decreasing* proportional band. With the nonlinear controller, amplitude *increases* with *increasing* deadband width.

The long period indicates that *reset* action has caused the loop gain to exceed 1.0. It is possible to eliminate this cycle by narrowing the deadband width or by increasing the deadband gain, to provide more control action at lower deviations.

Adjustment of the proportional band should be made for fast recovery from upsets that drive the measurement outside of the deadband. If the proportional band is too narrow, an excursion outside the deadband on one side will produce enough corrective action to drive the measurement out the other side, thereby creating a limit cycle larger than the deadband, but at the natural period.

Unsymmetrical Functions

The nonlinear function described is symmetrical, and is therefore only useful for control in the neutral region. If control is desired at some other point, for example, pH 3, nonlinear compensation is desirable but it must be asymmetric. For a simple titration curve, it is possible to remove the high-gain region from one side of the set point and adjust the deadband width so that its edge coincides with the knee of the curve.

To fit more complex curves, the deadband width may be automatically adjusted as a function of the controlled pH. Consider the adjustment of waste pickle liquor, consisting primarily of ferrous ions and either hydrochloric or sulfuric acid, to pH 10. This value is selected as effective in causing virtually complete precipitation of the iron as ferrous hydroxide. A titration curve for ferrous sulfate given in Fig. 3.7 illustrates the magnitude of the problem.

Breakpoints appear near pH 8 and again about pH 11—not equally distributed around the set point of 10. Furthermore, the slope of the titration curve below pH 8 is greater than that above

Special Components 199

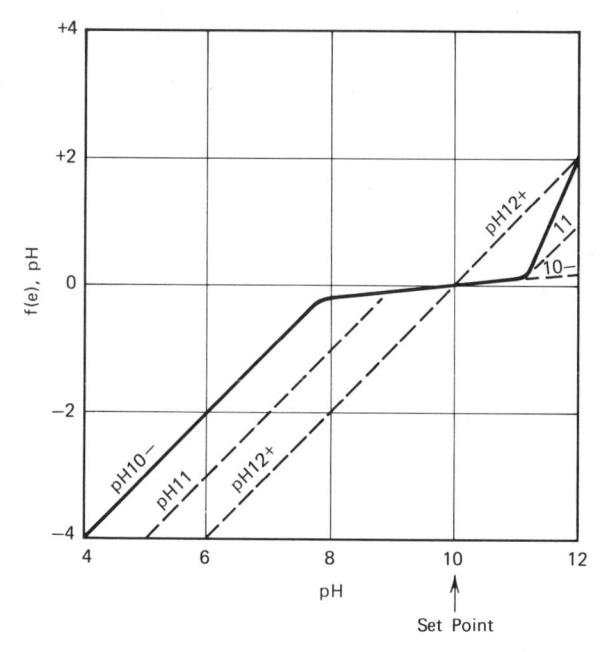

FIGURE 8.8. Changing the deadband width as a function of pH generates this unsymmetrical curve.

pH 11. The proper nonlinear function should then have a deadband between pH 8 and 11, with a higher gain above 11 than below 8.

This can be accomplished by making the deadband width change with pH but only over the range of pH 10 to 12. At pH 12, the deadband width should be zero, while at pH 10, it should be ± 2 pH units, as shown in Fig. 8.8. When the measurement is below the set point of 10, the width is -2 pH, giving a breakpoint at pH 8 with a gain of 1.0 below it. When the measurement is above 10, the deadband contracts one unit for every unit increase in pH. This develops a width of $+1$ pH, giving an upper breakpoint of pH 11 followed by a slope of 2.

A scaling device is used to take the pH measured over the full span of 4 to 12, and convert it to the limited span of 10 to 12 to drive the deadband width.

Sampled-Data Control

Some types of controllers may be transferred between automatic and manual operation bumplessly by a contact closure. Superior control can be achieved over dead-time dominant processes if such a controller can be operated in automatic periodically for short intervals separated by the dead time. Then corrective action imposed by the controller is allowed to appear completely in the controlled variable before more corrective action is applied. Only the integral mode is necessary, and if adjusted for a loop gain of 1, will produce optimum control. A loop gain of 2 in this system is required to produce uniform oscillation, and 1.5 gives quarter-amplitude damping.

Sampled-data control[3] is implemented simply by operating the auto-manual transfer switch of the controller with a repeat-cycle timer. Its principal disadvantage is that disturbances affecting the process while the controller is in manual will not be acted upon until the next cycle begins. Except for this problem, it is from two to four times as effective as a continuous controller on dead-time processes.

PERFORMANCE

Specifications on pH control systems, particularly in the area of waste treatment, are usually given as two hard limits within which control must be maintained. This is reasonable in that aquatic life of all forms is sensitive to both high and low pH. In cases where oxidation-reduction reactions are being carried out, usually only one pH limit applies. However, control too far on the safe side from this limit can increase the consumption of chemicals markedly, and also affect the accuracy of the redox measurement, most of which are pH sensitive. Similar rules apply to redox control—poor control (on the safe side) uses excess reagent which can also harm the environment.

It is one thing to mandate limits but quite another to control within those limits or even to estimate what will be required to achieve such control. Two situations must be considered:

1. Steady-state stability,
2. Dynamic response to upsets.

For the system described in Example 6.5, the use of a linear controller with a maximum proportional band of 300% resulted in a limit cycle of pH 5.4 to 8.6. Clearly this system would not meet specifications of pH 6 to 8 even without disturbances.

If the reagent flow were adjusted manually, it might be possible to bring the final pH between 6 and 8 in the steady state. But a slight change in influent flow or strength would drive it outside the limits. Between the too active controller which limit-cycles, and an inactive controller which cannot correct for disturbances, lie those whose effectiveness is more difficult to evaluate. They will eventually correct for disturbances, but fast enough to satisfy the process? To answer this question we must first analyze the response of the feedback loop.

Closed-Loop Response

To simplify the analysis, consider the composition-control loop whose block diagram appears in Fig. 8.9 to be linear. The series of calculations below will provide an estimate of the response of the effluent base concentration $c(\tau)$ to sinusoidal variations in acid load $q(\tau)$ occurring at a period of τ. Let the process consist simply of a dynamic response vector $g_1(\tau)$, exhibiting gain $G_1(\tau)$ and phase $\phi_1(\tau)$:

$$c(\tau) = g_1(\tau)[m(\tau) - q(\tau)]$$

The controller vector is $g_c(\tau)$, consisting of gain $G_c(\tau)$ and phase $\phi_c(\tau)$, and the valve gain is given as G_v:

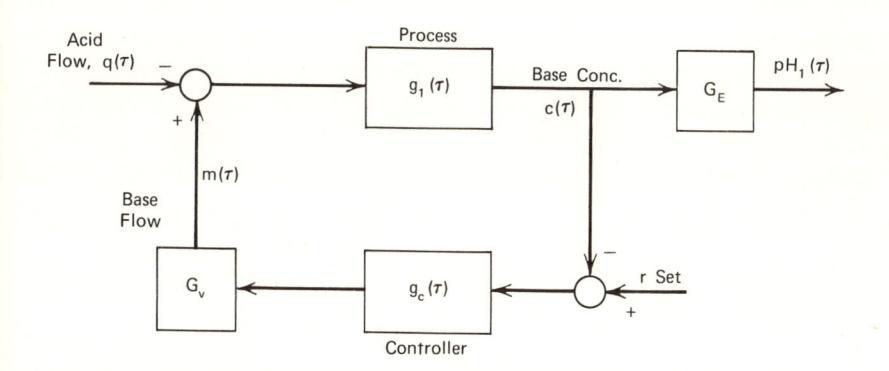

$$m(\tau) = G_v g_c(\tau)[r - c(\tau)]$$

FIGURE 8.9. A linear representation of the pH control loop.

Feedback Control Systems

Substituting for $m(\tau)$:

$$c(\tau) = g_1(\tau)\{G_v g_c(\tau)[r - c(\tau)] - q(\tau)\}$$

Solving for $c(\tau)$:

$$c(\tau) = \frac{g_1(\tau)[G_v g_c(\tau) r - q(\tau)]}{1 + g_1(\tau) G_v g_c(\tau)}$$

Since we are only interested in the response due to changes in load,

$$\frac{dc}{dq}(\tau) = -\frac{g_1(\tau)}{1 + g_1(\tau) G_v g_c(\tau)} = -\frac{1}{G_v g_c(\tau) + 1/g_1(\tau)} \quad (8.7)$$

From (8.7) it appears that increasing controller gain and decreasing process gain both favor control, and for stability reasons they occur together.

Bear in mind that $g_1(\tau)$ and $g_c(\tau)$ are dynamic terms—vectors whose gain and phase vary with the period of the imposed disturbance $dq(\tau)$. Qualitatively speaking, if the disturbance has a short period, $g_1(\tau)$ will tend to be low, giving but a slight response. If the period of the disturbance is extremely long, the reset component of $g_c(\tau)$ will give a very high gain, again resulting in only a slight response.

The response will be evaluated for disturbances over a range of periods τ, on both sides of the natural period of the control loop τ_o. The process will be defined as consisting of dead time τ_d plus a relatively large time constant τ_1, such that $\tau_o = 4\tau_d$. The dynamic gain of the process will be that of the time constant τ_1, with a phase angle of 90° for the time constant, plus a period-dependent amount for the dead time. At a period of τ_o, the phase angle for the dead time is $-90°$—at other periods it is adjusted proportionately. Specifically, at its natural period τ_o,

$$g_1(\tau_o) = \frac{\tau_o}{2\pi\tau_1 F} \angle -(90 + 90)$$

and at a period of τ,

$$g_1(\tau) = \frac{4\tau_d}{2\pi\tau_1 F}\left(\frac{\tau}{\tau_o}\right) \angle -\left(90 + \frac{90\tau_o}{\tau}\right) \quad (8.8)$$

The controller will be three mode, with a dynamic response related to the imposed period τ:

$$g_c(\tau) = \frac{100}{P}\sqrt{1+\left(\frac{2\pi D}{\tau}-\frac{\tau}{2\pi R}\right)^2} \angle \tan^{-1}\left(\frac{2\pi D}{\tau}-\frac{\tau}{2\pi R}\right)$$

If the controller is optimally adjusted to the period of the process, derivative (D) and reset (R) times will equal $\tau_o/2\pi$. Then,

$$g_c(\tau) = \frac{100}{P}\sqrt{1+\left(\frac{\tau_o}{\tau}-\frac{\tau}{\tau_o}\right)^2} \angle \tan^{-1}\left(\frac{\tau_o}{\tau}-\frac{\tau}{\tau_o}\right) \quad (8.9)$$

Furthermore, the loop gain at τ_o will be set at 0.5:

$$g_c(\tau_o)g_1(\tau_o)G_v = 0.5 \angle -180$$

At this point the phase contribution of the controller is zero, and the gain is simply

$$g_c(\tau_o) = \frac{100}{P}$$

Solving for $100/P$,

$$\frac{100}{P}\left(\frac{4\tau_d}{2\pi\tau_1 F}\angle -180\right)G_v = 0.5 \angle -180$$

$$\frac{100}{P} = \frac{0.5}{G_v}\left(\frac{2\pi\tau_1 F}{4\tau_d}\right) = \frac{0.78F}{G_v}\left(\frac{\tau_1}{\tau_d}\right) \quad (8.10)$$

Using values of $g_1(\tau)$, $g_c(\tau)$, and $100/P$ calculated by (8.8) to (8.10), Table 8.1 was prepared following (8.7). Notice that the valve gain does not appear in the right-hand column. This is because, whatever its value, the controller gain has to be adjusted to compensate for it, so the valve could have been left out of the analysis as the transmitter gain was. In the region of τ/τ_o between 0.5 and 2, the process is more sensitive to disturbances with control than without, since without control $(dc/dq)(\tau)$ is simply $G_1(\tau)$. For shorter periods,

204 Feedback Control Systems

Table 8.1 Sensitivity of the Closed Loop to Disturbances

τ/τ_o	$g_1(\tau)$ $G_1(\tau)$	$\phi_1(\tau)$ (degrees)	$g_c(\tau)$ $G_c(\tau)$	$\phi_c(\tau)$ (degrees)	$\dfrac{dc}{dq}(\tau)$
10	$6.4\,\tau_d/\tau_1 F$	-99	$7.8\,F\tau_1/\tau_d G_v$	-84	$-0.13\,\tau_d/\tau_1 F$
4	2.56	-112	3.02	-75	-0.38
2	1.27	-135	1.40	-56	-1.63
1	0.64	-180	0.78	0	-1.28
0.5	0.32	-270	1.40	$+56$	-0.58
0.25	0.16	-360	3.02	$+75$	-0.13

$(dc/dq)(\tau)$ approaches $G_1(\tau)$, whereas for longer periods, it approaches $1/G_c(\tau)G_v$:

$$\text{for } \frac{\tau}{\tau_o}<0.5, \quad \frac{dc}{dq}(\tau)\to G_1(\tau)=\frac{\tau}{2\pi\tau_1 F}=0.64\frac{\tau}{\tau_o}\left(\frac{\tau_d}{\tau_1 F}\right) \quad (8.11)$$

$$\text{for } \frac{\tau}{\tau_o}>2, \quad \frac{dc}{dq}(\tau)\to\frac{1}{G_c(\tau)G_v} \quad (8.12)$$

At the longer periods, reset is the dominant control mode, such that

$$\text{for } \frac{\tau}{\tau_o}>2, \quad G_c(\tau)\to\frac{100}{P}\left(\frac{\tau}{2\pi R}\right)$$

Substituting for P and R yields

$$G_c(\tau)\to\frac{0.78 F}{G_v}\left(\frac{\tau_1}{\tau_d}\right)\left(\frac{\tau}{\tau_o}\right)$$

Then,

$$\text{for } \frac{\tau}{\tau_o}>2, \quad \frac{dc}{dq}(\tau)\to 1.28\frac{\tau_o}{\tau}\left(\frac{\tau_d}{\tau_1 F}\right) \quad (8.13)$$

Let the coefficient of $\tau_d/\tau_1 F$ appearing in the last column of Table 8.1 and in (8.11) and (8.13) be designated $y(\tau)$. This period-dependent component is plotted against τ/τ_o in Fig. 8.10.

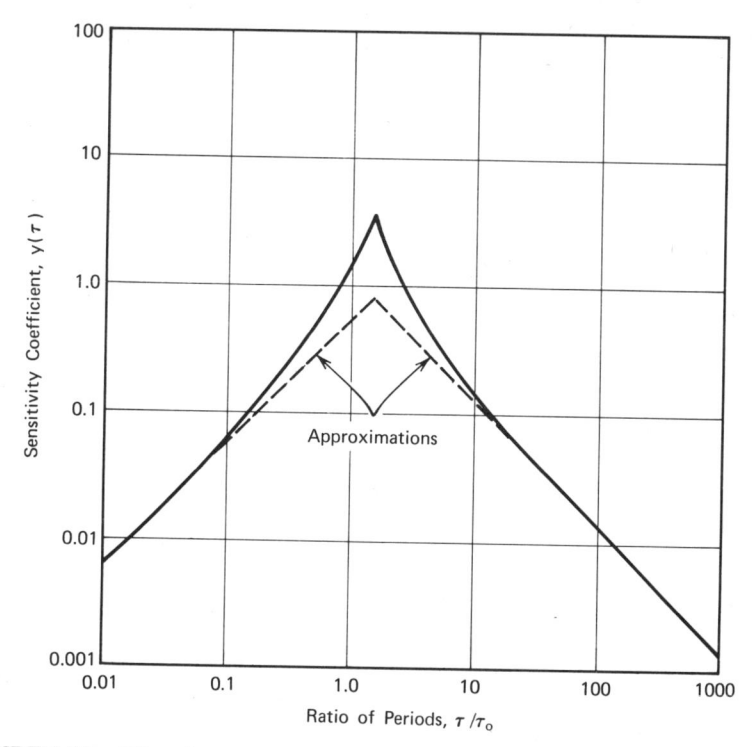

FIGURE 8.10. Disturbances occurrring near the natural period of the loop cause the most trouble.

To be meaningful to people with pH control problems, $(dc/dq)(\tau)$ should be in terms of pH-units variation in the effluent versus pH-units variation in the influent. Base concentration change $dc(\tau)$ may be converted into $d\text{pH}(\tau)$ using electrode gain G_E. To convert disturbance $dq(\tau)$ into an equivalent change in pH at constant flow F requires differentiating. Let the influent acid concentration be x_A:

$$x_A = \frac{q}{F}$$

Then,

$$\text{pH}_0 = -\log x_A$$

and,

$$\frac{d\text{pH}_0}{dx_A} = -\frac{\log e}{x_A} = -\frac{0.434}{10^{-\text{pH}_0}}$$

Finally,

$$\frac{dpH_0}{dq} = -\frac{0.434}{F(10^{-pH_0})} \qquad (8.14)$$

Electrode gain may be estimated from an appropriate titration curve. For the case of strong agents only, its maximum value is the slope of the titration curve at neutrality, that is, 2.17×10^6 pH/N, as determined in Chapter 3.

Combining the factors above yields

$$\frac{dpH_1}{dpH_0}(\tau) = \frac{(dc/dq)(\tau)(dpH_1/dc)}{dpH_0/dq}$$

Let $(dc/dq)(\tau)$ be represented by $-y(\tau)\tau_d/\tau_1 F$. Then,

$$\frac{dpH_1}{dpH_0}(\tau) = -\frac{y(\tau)\tau_d}{F\tau_1}(2.17 \times 10^6)\left(-\frac{F \times 10^{-pH_0}}{0.434}\right)$$

Combining terms yields

$$\frac{dpH_1}{dpH_0}(\tau) = 0.5 \times 10^{7-pH_0} \frac{y(\tau)\tau_d}{\tau_1} \qquad (8.15)$$

Equation (8.15) concludes that the sensitivity of the process to disturbances varies with the mixing (τ_d/τ_1), the period of the upset (τ), and the distance between the influent pH and the control point ($7-pH_0$).

Example 8.2

Estimate the sensitivity of a pH control loop to cyclic variations in influent pH at a 10-min period, assuming a strong acid-base reaction, with the influent at pH 2. Let $\tau_d = 15$ sec and $\tau_1 = 3$ min.

$$\tau_0 = 4\tau_d = 60 \text{ sec} = 1.0 \text{ min}$$

$$\tau/\tau_0 = 10$$

$$y(10) = 0.13$$

$$\frac{d\text{pH}_1}{d\text{pH}_0}(10) = 0.5 \times 10^{7-2}(0.13)\left(\frac{15}{60 \times 3}\right)$$

$$\frac{d\text{pH}_1}{d\text{pH}_0}(10) = 540$$

Thus a change in influent pH of about 0.002 units could change the effluent pH one unit.

Usually disturbances are not periodic but random variations comprised of a combination of periodic inputs. Step changes can occur in some processes, but usually not in waste streams. The sensitivity analysis just concluded is intended to demonstrate

1. the effectiveness of control,
2. the importance of keeping dead time small, and
3. the need for additional means for reducing the sensitivity to disturbances.

Buffering can, of course, reduce the sensitivity significantly. In addition, a wider tolerance in effluent pH can also help, by reducing the electrode gain.

Example 8.3

For the same process as Example 8.2 estimate the change in influent pH required to change the effluent pH between 5 and 9. The electrode gain is $(\text{pH}_1 - 7)/c = 2 \text{ pH}/10^{-5} \, N$ or $2 \times 10^5 \text{ pH}/N$, compared to 2.17×10^6 used above. Therefore

$$\frac{d\text{pH}_1}{d\text{pH}_0}(10) = 540\left(\frac{2 \times 10^5}{2.17 \times 10^6}\right) = 50$$

$$d\text{pH}_0(10) = \frac{9-5}{50} = 0.08$$

The greatest influence over sensitivity is the difference between influent pH and the control point—one unit closer can mean one decade reduction in sensitivity. Where estimates indicate sensitivity to be too great to meet specifications, other means of reducing it must be employed. The remedy most readily applied is a smoothing tank, described in Chapter 7.

The Effect of Smoothing

An additional vessel, even if unmixed, can dynamically reduce sensitivity to disturbances. The dynamic gain of the second vessel to variations in acid or base concentration is

$$G_2(\tau) = \frac{1}{\sqrt{1 + (2\pi\tau_2/\tau)^2}} \quad (8.16)$$

The time constant τ_2 is less than the residence time by the dead time which, for an unmixed vessel, may be about one-third of the residence time. So the absence of mixing in the second vessel makes it only 65 to 70% effective, but it does not hurt control as much as it would in the vessel with the feedback loop. The second vessel may be located either upstream or downstream of the feedback loop.

Example 8.4

Using the information given in Example 8.2, estimate the sensitivity reduction afforded by adding a smoothing vessel with a 10-min residence time and a 3-min dead time.

$$G_2(10) = \frac{1}{\sqrt{1 + (2\pi 7/10)^2}} = 0.22$$

Because of the nature of the smoothing vessel, it is most effective on rapidly varying influent disturbances or in attenuating limit cycles in the control loop. If located ahead of the neutralization vessel, it can only moderate influent disturbances, while if located downstream it can do both jobs.

The high sensitivity of the pH measurement in the neutral range prevents the controller from being especially active there. Only deviations beyond this range will then elicit effective control action —yet these deviations usually exceed environmental standards and the waste cannot be discharged in that condition. But a downstream smoothing tank can attenuate these deviations if control response is sufficiently rapid in comparison to the residence time of the tank, and thereby allow continuous discharge.

Multiple-Stage Neutralization

Sensitivity to disturbances can also be reduced by neutralizing in two stages, each with a control loop. Instead of bringing the influent from pH 2 to 7 in one vessel, for example, it could be raised to pH 4 in the first stage and then to 7 in the second.

In the first stage,

$$\frac{d\text{pH}_1}{d\text{pH}_0}(\tau) = -\frac{y_1(\tau)\tau_{d1}}{F\tau_1}\left(\frac{d\text{pH}_1}{dc_1}\right)\left(-\frac{F \times 10^{-\text{pH}_0}}{0.434}\right)$$

Since there is a deficiency of base leaving stage 1, c is negative:

$$\frac{d\text{pH}_1}{dc_1} = -\frac{0.434}{c_1} = \frac{0.434}{10^{-\text{pH}_1}}$$

Then

$$\frac{d\text{pH}_1}{d\text{pH}_0}(\tau) = 10^{\text{pH}_1 - \text{pH}_0}\left[\frac{y_1(\tau)\tau_{d1}}{\tau_1}\right] \qquad (8.17)$$

Equation (8.15) is then restated for pH_2:

$$\frac{d\text{pH}_2}{d\text{pH}_1}(\tau) = 0.5 \times 10^{7-\text{pH}_1}\left[\frac{y_2(\tau)\tau_{d2}}{\tau_2}\right] \qquad (8.18)$$

Combining the two expressions,

$$\frac{d\text{pH}_2}{d\text{pH}_0}(\tau) = 0.5 \times 10^{7-\text{pH}_0}\left[\frac{y_1(\tau)\tau_{d1}}{\tau_1}\right]\left[\frac{y_2(\tau)\tau_{d2}}{\tau_2}\right] \qquad (8.19)$$

The reduction in sensitivity achieved by the second stage is simply $y_2(\tau)\tau_{d2}/\tau_2$.

One precaution must be taken when selecting vessel size. If the second vessel is made the same size as the first, it will be sensitive to variations in pH leaving the first vessel at the natural period of the first loop. Refer to Table 8.1 and let τ become the period of the first loop and τ_o the period of the second. If the period of the first loop is one-fourth that of the second, this sensitivity is reduced tenfold. So the second vessel should be four times as large as the first.

A tenfold reduction in sensitivity could also be achieved if the first vessel were *ten* times the size of the second, but the former approach seems the more practical.

Example 8.5

Let the stream described in Example 8.2 be neutralized to pH 4 in the existing vessel and to pH 7 in a second stage of 10-min residence time and 1-min dead time. Estimate the reduction in sensitivity afforded by the second vessel:

$$\tau_{o2} = 4\tau_{d2} = 4 \text{ min}, \quad \frac{\tau}{\tau_{o2}} = 2.5, \quad y_2(10) = 1.2$$

$$\frac{y_2(10)\tau_{d2}}{\tau_2} = \frac{(1.2)(1.0)}{9} = 0.132$$

The second stage is almost twice as effective as the same size smoothing tank in Example 8.4.

In summary, waste-neutralization processes are seen to be very sensitive to disturbances, even with the nonlinearity completely compensated and the controller optimally adjusted. Fortunately some buffering exists in virtually every process, or few plants would be controlled acceptably. Smoothing vessels and multiple stages are helpful in reducing this sensitivity but only by one order of magnitude or so, unless outsized vessels are chosen. Another method of reducing the sensitivity to disturbances is feedforward control—its application is described in depth in the next chapter.

BATCH PROCESSES

Special consideration must be given to those processes which are operated batchwise and those which, though continuous, are interrupted daily or weekly. Batch processing has some considerable advantages, the most important of which is an ability to retain an effluent until its quality meets specifications before release. When treating extremely toxic wastes, this safeguard is almost mandatory.

Economic incentives lean toward batch processes either when the plant production units are operated batchwise, or when flow rates are low, typically less than 100 gpm. A single vessel can be used to add caustic, oxidize cyanide to cyanate, adjust pH, destroy cyanate, and neutralize batchwise, where two or three would be required for continuous operation. In many cases, however, duplicate batch vessels have to be used, with one being filled while in the other, treatment is proceeding.

Batch Composition Control

In a continuous process, the controlled variable is to be maintained at the set point by balancing the manipulated variable against the load. In a batch process, however, the only time the measurement is at the set point is at the end of the run. At this point reagent addition should be over —otherwise the controlled variable will overshoot its mark.

If a controller with reset action is used on a batch process, it will drive the valve fully open and keep it there until the controlled variable crosses the set point. Proportional action is desirable without reset; however, the controller must be adjusted so that the valve is shut when the error is zero.

Overshoot is still possible, however, since some delay always exists between the motion of the valve and the resulting response in the controlled variable. Once an overshoot occurs, it cannot ordinarily be corrected. If adding acid to reduce pH to 7, for example, a base would have to be added to correct an overshoot, and none may be available. To preclude an overshoot, then, the valve must be closed before the controlled variable reaches the set point. This can be done with derivative action, by setting the derivative time at or near the delay in response of the controlled variable to the valve. Too little derivative will still permit some overshoot, while too much will close the valve prematurely. In the latter case, the controlled variable will fall short of the set point, but only momentarily, for the valve will then open again. Figure 8.11 displays a typical record of pH versus time for a batch process controlled by the system shown in Fig. 8.12. A flow-through electrode assembly is used to avoid uncovering the electrodes when the tank is empty.

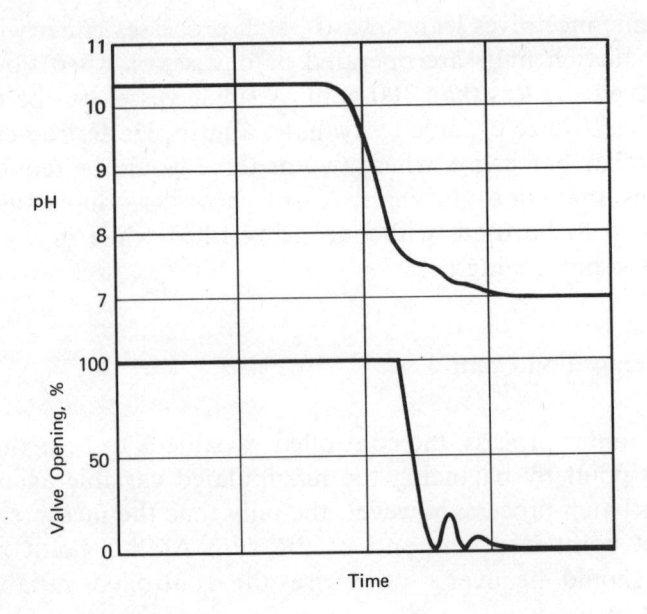

FIGURE 8.11. Derivative can shut the valve before the pH measurement reaches the set point.

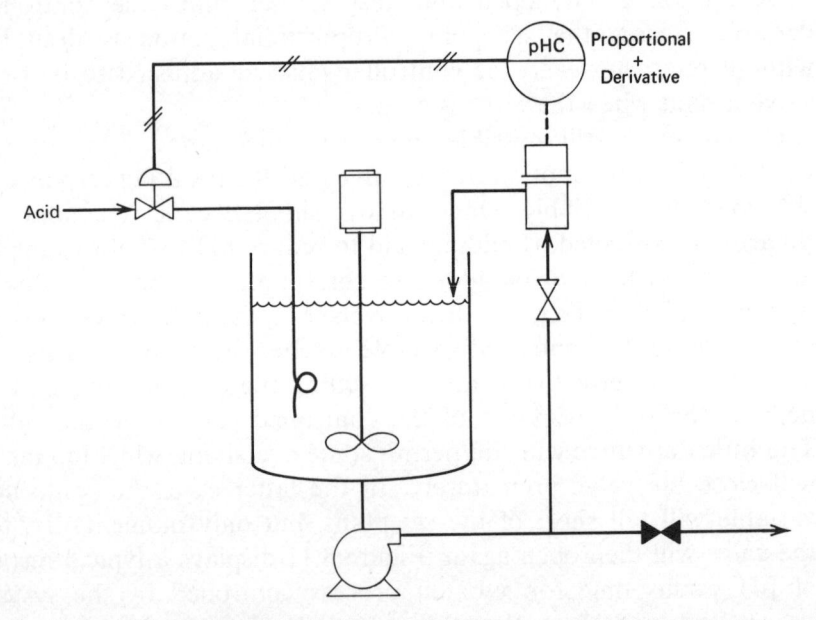

FIGURE 8.12. Proportional-plus-derivative control is necessary to avoid overshoot in batch processes.

On-off control may be used in situations where electrode gain is relatively low, as when adjusting pH to 2.5 prior to reducing chromate. Reagent would be added through an on-off valve at a constant rate, until the measurement reaches the set point, when it would be shut off altogether.

For control in regions of moderate sensitivity, the reagent valve could remain on till some value short of the set point was reached. Then, the valve could be operated by a repeat-cycle timer, allowing it to remain open for only a fraction of the cycle time. This action provides a more gradual approach to the set point. If the cycle time is set longer than the delay in response of the controlled variable to a change in reagent flow, the loop will be stable. However, more time will be required to reach the set point from a given starting condition than with properly adjusted proportional control.

Interrupted Continuous Operation

All the rules given for effective control of continuous process apply here, with the added requirement of automatic startup. When a continuous process is shut down, the controlled variable no longer responds to a change in controller output. As a result, a conven-

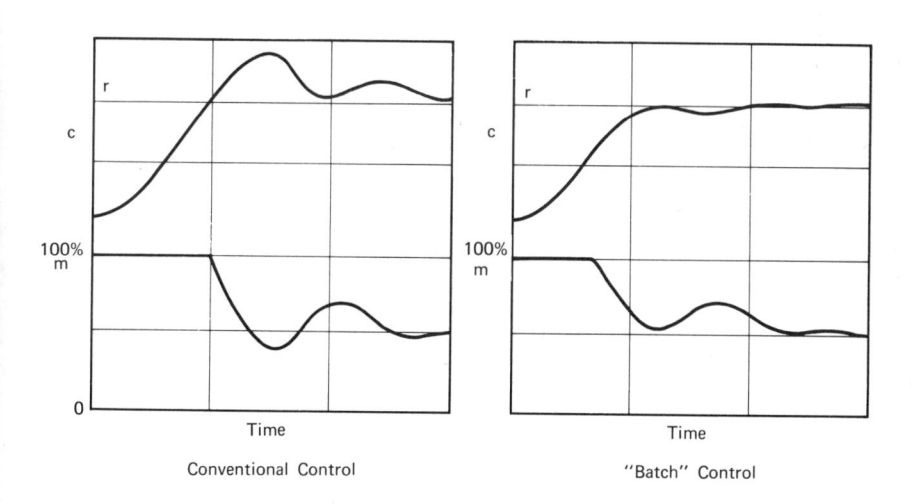

FIGURE 8.13. Logic especially designed for batch operation can avoid overshoot following startup.

tional controller will drive its output to one extreme or the other, trying to correct a condition beyond its influence.

The danger in this condition is that upon startup, that extreme value of controller output will be applied to the reagent flow, until the controlled variable crosses the set point. Thus overshoot will invariably result, and several cycles may be required before the controlled variable is within acceptable limits. The problem here is, at least during shutdown, the same as that of a pure batch process. And the solution is similar—reset action must be inhibited during shutdown.

To disable reset during shutdown and yet enable it during operation requires a logic attachment to the controller specifically designed for batch processes. If, during shutdown, the controller tends to raise its output to full scale, the logic can disable the reset circuit when full-scale output is reached. This action converts the controller to proportional (or proportional plus derivative) until its output again falls below full scale at startup. Figure 8.13 illustrates the improvement achieved.

REFERENCES

1. F. G. Shinskey, *Process-Control Systems,* McGraw-Hill Book Company, New York, 1967, pp. 128–131.
2. Hammel-Dahl Division of ITT Corporation, MOD[(T)] Modulating Ball Valve.
3. F. G. Shinskey, *op. cit.*, pp. 110–117.

9

FEEDFORWARD AND ADAPTIVE CONTROL

Hopefully the preceding eight chapters have demonstrated how difficult the control of ion concentration can be, particularly in treating *mixtures* of plant wastes. Not only is the process extremely sensitive to upsets—which can be severe and wide ranging—but its response characteristics are also quite variable. Feedback control is of limited effectiveness then, not only because of the frequency and magnitude of the *disturbances* encountered but also because the controller is seldom adjusted properly for the *conditions* which prevail.

This last chapter is devoted to resolving this twofold dilemma—more powerful techniques must be used to improve control beyond the capabilities of simple feedback loops. To reduce the sensitivity of the process to upsets we use feedforward control; adaptive loops are devised to adjust the control system for variations in the parameters of the process. Very often these tools will be used together in order to maximize their individual effectiveness. In fact they are in practice so complementary that they borrow one another's properties in achieving their objectives.

FEEDFORWARD CONTROL

Feedforward control[1] is that technique by which variations in load are converted directly into changes in the manipulated variable to cancel their effect on the process. It is used extensively by each one of us in our everyday actions. In driving a car, for example, the driver adjusts his speed and steering based on impending situations —hills, curves, intersections, and so on—*before* a collision occurs. Preventive medicine and preventive maintenance are both examples of feedforward control, as is withholding tax. By contrast, corrective medicine and repairs are feedback modes of operating—no corrective action is taken until some "damage" becomes apparent.

In theory, feedforward is capable of perfect regulation, since it does not depend on a deviation of the controlled variable for its stimulus. In practice, however, it is impossible to position the manipulated variable with absolute accuracy with respect to the exact load on the process at any given time. However, its response can be virtually instantaneous, relative to the dead time and capacity found in every feedback loop, making it a useful adjunct to feedback even when limited accuracy precludes its use alone. If, for example, the feedforward calculation is only accurate to ±20%, by reducing the contribution of feedback to only 20% a control improvement of 5:1 will be realized.

In a feedforward pIon control system, the manipulated flow of reagent is calculated to balance the flow of the companion ion entering the process. Consider the case of acid to be neutralized by an equivalent amount of caustic. If the flow F_A and normality x_A of acid are measurable, the equivalent flow of base F_B of known concentration x_B can be calculated:

$$F_B = \frac{F_A x_A}{x_B} \tag{9.1}$$

This calculation is easy enough to make, but obtaining the input information and manipulating the output present problems.

Measuring Flow

There are three principal methods of measuring flow in ion-control systems: magnetic flowmeters, orifice meters, and open-channel meters. The magnetic flowmeter transmits a signal proportional to the velocity of the fluid passing through its magnetic field. It is essentially a pipe surrounded by a set of coils generating an alternating-current field. The flowing path is therefore obstructionless. Electrodes contacting the flowing fluid are insulated from the pipe with a nonconducting liner of abrasion-resistant plastic or rubber. The fluid itself must be conducting, but only to the extent of 2 μmho/cm or so.

The magnetic flowmeter produces a signal linear with flow and is accurate to about 1% of full span over a range of 10:1. Its principal problems are sensitivity to foreign materials which may interfere with the measurement. Entrained air, or a partly empty flowmeter, can cause severe errors and erratic behavior.

Orifice flowmeters should be sufficiently familiar that their general properties need only be touched upon here. For a guide to making this type of measurement, Reference 2 is recommended. Flow F through an orifice of area A varies with the square root of differential pressure, h:

$$F = kA\sqrt{h} \qquad (9.2)$$

Nozzles and venturi meters have the same general characteristics as orifice meters, but are of more streamlined construction, resulting in less permanent loss in head. A $\pm 1\%$ accuracy is obtained only over a 3:1 or 4:1 range with orifice-type meters.

Open-channel flow measurements are common in waste treatment in that large volumes of water can be easily measured with little loss in head. A weir with a rectangular or triangular notch may be used, across which a head develops proportional to the flow rate. The relationship between head and flow is similar to that of the orifice, except that the area through which flow passes also changes with the head. For rectangular weirs of width w,

$$A = wh \qquad (9.3)$$

and for triangular weirs of apex angle α,

$$A = h^2 \tan \frac{\alpha}{2} \qquad (9.4)$$

Combining these two relationships with (9.2) yields

$$F = kwh^{3/2} \qquad (9.5)$$

218 Feedforward and Adaptive Control

for rectangular weirs, and

$$F = k\left(\tan\frac{\alpha}{2}\right)h^{5/2} \tag{9.6}$$

for triangular weirs. The characteristics of all three of these flowmeters appear in Fig. 9.1.

Parshall flumes are contoured channels developing a head proportional to flow in much the same manner as a rectangular weir, but with less permanent head loss. Thus they compare to rectangular weirs in the same way that venturis relate to orifice meters.

Determining Normality

Normality measurements are not so easy to make. The only exact way to determine the amount of reagent required to neutralize a

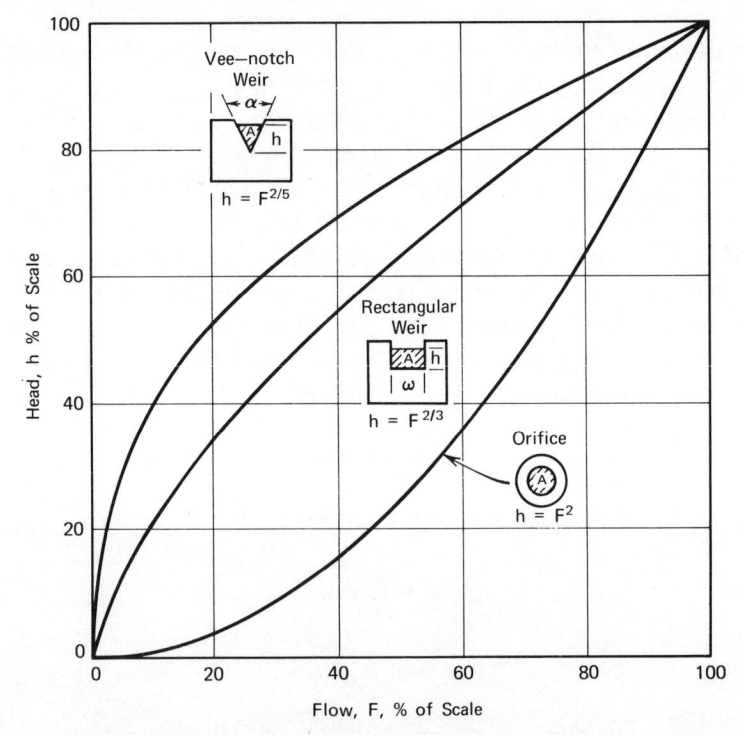

FIGURE 9.1. Characteristics of common nonlinear flowmeters.

given waste is to run an actual titration. Automatic titrators are available from several manufacturers, but they all operate batchwise, analyzing samples drawn at discrete intervals of 1 min at best. Since the period of most pH feedback loops approaches this, no advantage is gained using a batch titrator for feedforward control.

A continuous titrator would be, in fact, a small-scale model of the actual process. By continuously neutralizing the waste actually entering the facility, it would be duplicating the functions of the control system with all its attendant problems. It would require feedback control with similar rangeability and stability but much faster dynamic response if it is to provide information useful to the main control system. Since the dynamic response of the main control loop is limited by mixing, electrode condition, and valve response, the titrator must have much faster mixing—the other elements are not likely to be improved upon. This would tend to restrict the use of the continuous titrator to processes whose mixing is poor and cannot be improved due to physical constraints.

A simple pH measurement of the untreated waste will indicate the normality by (3.5) if only strong agents are in the solution:

$$x_A - x_B = 10^{-pH} - 10^{pH-14} \tag{3.5}$$

For feedforward control purposes we are interested only in acid content below pH 6 and base content above pH 8, so that (3.5) may be simplified into

$$x_A = 10^{-pH} \quad \text{below} \quad pH \ 6 \tag{9.7A}$$

and

$$x_B = 10^{pH-14} \quad \text{above} \quad pH \ 8 \tag{9.7B}$$

Unfortunately, however, the presence of variable quantities of any weak acids and bases will alter these simple relationships, as indicated by (3.11) and (3.13). For a mixture of a weak acid, strong acid, and strong base, the sum $x_A - x_B + x_C$ must be neutralized where x_C is the normality of the weak component. Using (3.11) the relationship between $x_A - x_B$ and pH may be derived for constant x_C:

$$x_A - x_B = 10^{-pH} - \frac{x_C}{1 + 10^{pK_c - pH}} \tag{9.8A}$$

220 Feedforward and Adaptive Control

For a condition of constant $x_A - x_B$, the relationship between x_C and pH may be derived by rearrangement:

$$x_C = (10^{-pH} - x_A + x_B)(1 + 10^{pK_c - pH}) \qquad (9.9A)$$

These two relationships apply only below pH 6, where 10^{14-pH} is insignificant. Similar equations may be derived for systems with weak bases above pH 8:

$$x_B - x_A = 10^{pH-14} - \frac{x_D}{1 + 10^{pK_D + pH - 14}} \qquad (9.8B)$$

$$x_D = (10^{pH-14} - x_B + x_A)(1 + 10^{pK_D + pH - 14}) \qquad (9.9B)$$

A variety of curves of log-normality versus pH may be obtained depending on whether the weak or the strong components are varying. Figure 9.2 has been prepared to illustrate several of these curves representing different mixtures of acetic acid (x_C), hydro-

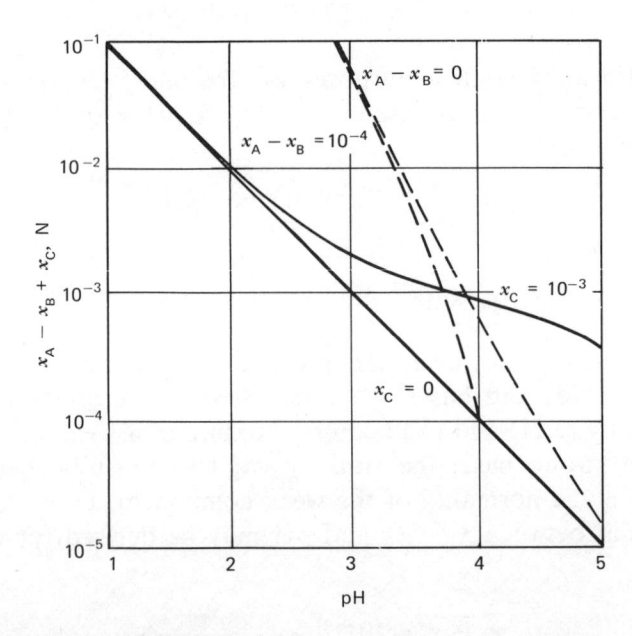

FIGURE 9.2. The relationship between pH and normality depends on which of the components is changing.

chloric acid (x_A), and caustic (x_B). All curves ride above the fundamental strong-acid line $(x_C=0)$ whose slope is 1.0. When the amount of strong agents varies with fixed x_C, the curve has a relatively flat central portion whose slope is less than 1.0, varying inversely with x_C; at low pH values the curve approaches the strong-acid line while at higher pH values it turns parallel to the line.

The pH of a single, monoprotic weak agent changes approximately one-half unit per decade change in concentration. Consequently the slope of the line with $x_A - x_B = 0$ in Fig. 9.2 is 2.0. However, with a fixed amount of strong agent in solution ($x_A - x_B = 10^{-4}$), the pH versus log-concentration curve departs from that of the pure weak-acid line to terminate at the strong-acid line. Its slope is at all points greater than 2.0.

Multiprotic agents or mixtures of weak agents complicate the pH versus log-concentration curve even more. The point to be made here is that pH is uniquely related to normality only with a single component in solution. For any mixture, it is only an approximation, and possibly a poor one—yet it is usually the only indication available.

Narrow-Range Systems

The distinction between narrow- and wide-range systems here is based on the required rangeability of the manipulated reagent flow. If the demand for reagent changes only over a 4:1 range, it may be accurately added by adjusting the set point of a flow controller whose measurement is the differential pressure across an orifice. Flow ranges down to 0.01 gpm are available with an orifice integrally attached to the differential-pressure transmitter. Since the differential pressure varies as flow-square, however, its square root must be extracted to provide a linear flow signal. Figure 9.3 illustrates a narrow-range feedforward system using orifice meters for both agents.

The ratio between ammonia and nitric acid in the figure is adjusted by the pH controller to form a neutral product. Should the *flow* of ammonia change, the flow of nitric acid will respond proportionally, so that product pH will not change appreciably. Should the *concentration* of the ammonia feed change, the pH will

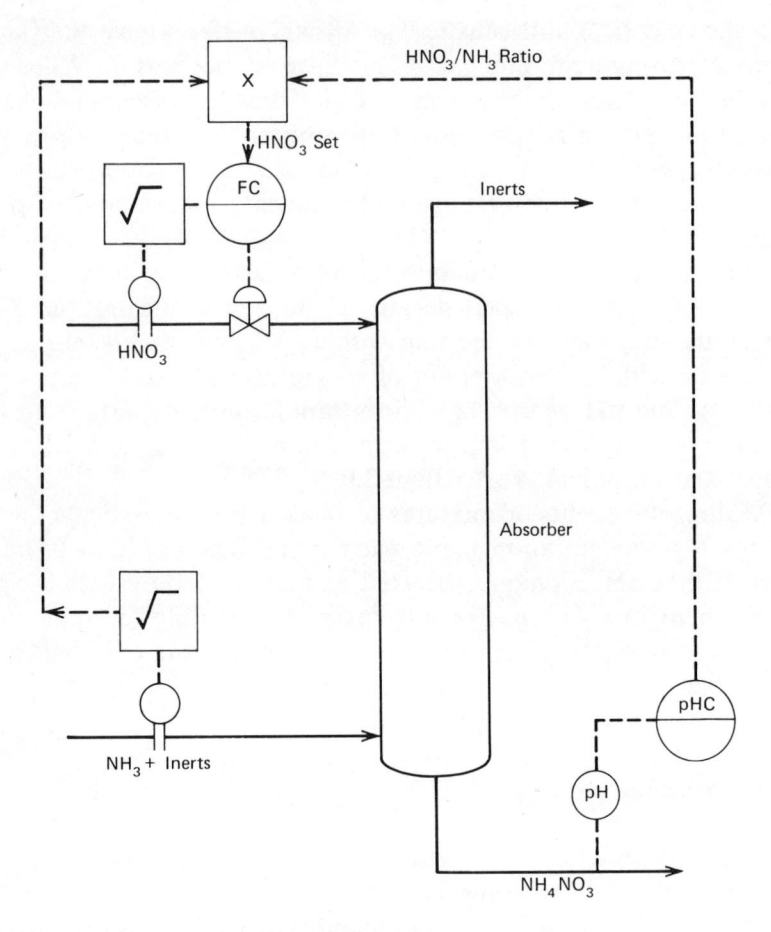

FIGURE 9.3. Narrow-range feedforward systems such as this are used for processes with nearly constant feed compositions.

deviate from the set point until the feedback controller readjusts the ratio of $HNO_3 : NH_3$ to the correct value for the new conditions. Narrow-range systems such as this are used primarily for chemical reactions where feedstocks are fairly consistent in quality and throughput is restricted to definite limits.

The magnetic flowmeters described earlier can also be used for reagent flow control extending the rangeability of the system to 10:1. They are available in flow ranges down to 0.1 gpm.

A preferred reagent-delivery device for linear narrow-range systems is the metering pump, with about 20:1 turndown. If both remotely adjustable stroke and speed are available, the feedforward flow signal and the feedback controller output may be combined without a multiplier. If not, the two signals are multiplied as when manipulating a linear flow-control set point or the position of a linear valve. Both arrangements are shown in Fig. 9.4.

A feedforward normality input may be added to the system if a reasonable relationship with pH exists. However, a function generator is necessary to convert the pH measurement to normality. Due to the limited rangeability of the metering pump, only about one decade of normality can be covered. Figure 9.5 shows how the pH measurement fits into the control system, with inserts describing the shape of the required nonlinear functions. Where head flowmeters are used, functions which are the inverse of those shown in Fig. 9.1 are needed to produce a linear flow signal.

Wide-Range Systems

Most waste-treatment systems require a capability of delivering reagent beyond a 20:1 range. In these cases, the only reasonable

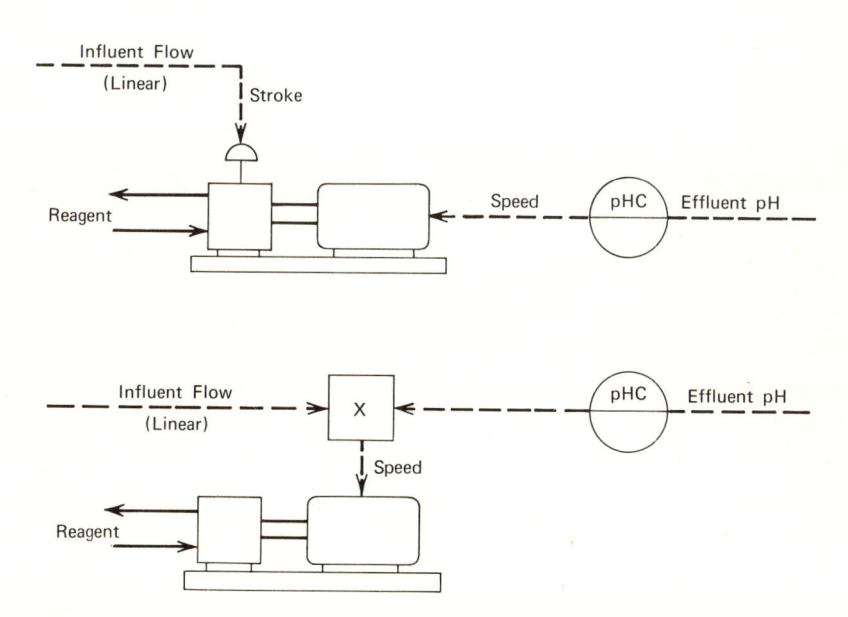

FIGURE 9.4. A metering pump with a remotely adjustable stroke as well as speed can accommodate a feedforward loop without a multiplier.

224 Feedforward and Adaptive Control

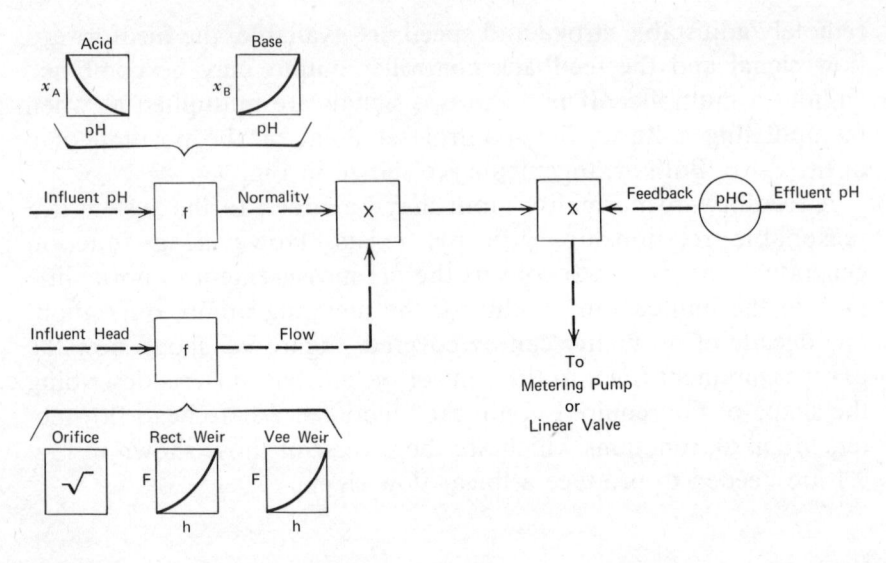

FIGURE 9.5. Function generators are required to linearize flow from a head measurement and to convert pH to normality.

choice is the equal-percentage valve for a rangeability up to 50:1, or two properly sequenced equal-percentage valves if 1000:1 range is required. (Methods for sequencing and the use of a separate "trim" controller to manipulate a third valve—thereby extending rangeability beyond 10,000:1—were described in Chapter 8.) However, this leaves the manipulated variable in an exponential form, following (6.18):

$$f = R^{m-1} \qquad (6.18)$$

where f is the fractional reagent flow, R is the rangeability of the valve(s), and m is the fractional signal to the valve(s).

Converting (6.18) and (9.7) to logarithms will reveal how this valve characteristic is compatible with the pH-normality relationship:

$$\log f = (m-1)\log R \qquad (9.10)$$

$$\log x_A = -\text{pH} \qquad (9.11\text{A})$$

$$\log x_B = \text{pH} - 14 \qquad (9.11\text{B})$$

Examine the logarithmic solution to the feedforward control

equation (9.1) for a waste acid:

$$\log F_B = \log F_A + \log x_A - \log x_B \qquad (9.12)$$

The actual reagent flow F_B is the product of the fractional flow f_B and the valve capacity, B; the waste flow F_A is similarly related to flowmeter range A, and flow signal f_A:

$$\log F_B = \log f_B + \log B \qquad (9.13)$$

$$\log F_A = \log f_A + \log A \qquad (9.14)$$

Combining the last four equations yields

$$(m-1)\log R + \log B = \log f_A + \log A - \text{pH} - \log x_B$$

Solving for the signal to the valves:

$$m = 1 + \frac{\log f_A - \text{pH} - \log Bx_B/A}{\log R} \qquad (9.15A)$$

The feedforward equation for acid neutralizing a basic waste is similar:

$$m = 1 + \frac{\log f_B + \text{pH} - 14 - \log Ax_A/B}{\log R} \qquad (9.15B)$$

where A is the capacity of the acid reagent valves.

Example 9.1

Consider neutralizing 1000 gpm of acid waste varying in pH from 2 to 5 with 10% caustic; two sequenced valves are used giving a rangeability of 1000:1, with a maximum flow of 5 gpm. Solve (9.15A), evaluating all the constants:

$$\frac{Bx_B}{A} = \frac{(5 \text{ gpm})(2.75\ N)}{1000 \text{ gpm}} = 0.01375\ N$$

$$\log \frac{Bx_B}{A} = -1.86$$

$$\log R = 3$$

$$m = 1 + 0.33(\log f_A + 1.86 - \text{pH})$$

226 Feedforward and Adaptive Control

At full-scale flow, $\log f_A = 0$; then the larger valve would be fully open at pH 1.86, with the smaller valve closed at pH 4.86.

Because f_A is a positive number less than 1.0, $\log f_A$ is always negative. To avoid working with negative numbers, the group $(1 + \log f / \log \mathcal{R})$ is broken out of (9.15) as a positive number. The shape of this function of flow is given versus flow in Fig. 9.6 for two values of rangeability. Observe that the characteristic curve of a Vee-notch weir is a reasonable model of these logarithmic curves, particularly for moderate rangeability. This, then, is the type of flowmeter most suited for a feedforward input with wide-range systems. A linear flow signal requires much more characterization, and an orifice differential requires so much gain, that with its limited rangeability, it is not worth using as a feedforward input.

A flowmeter with a very desirable characteristic is *another* equal-

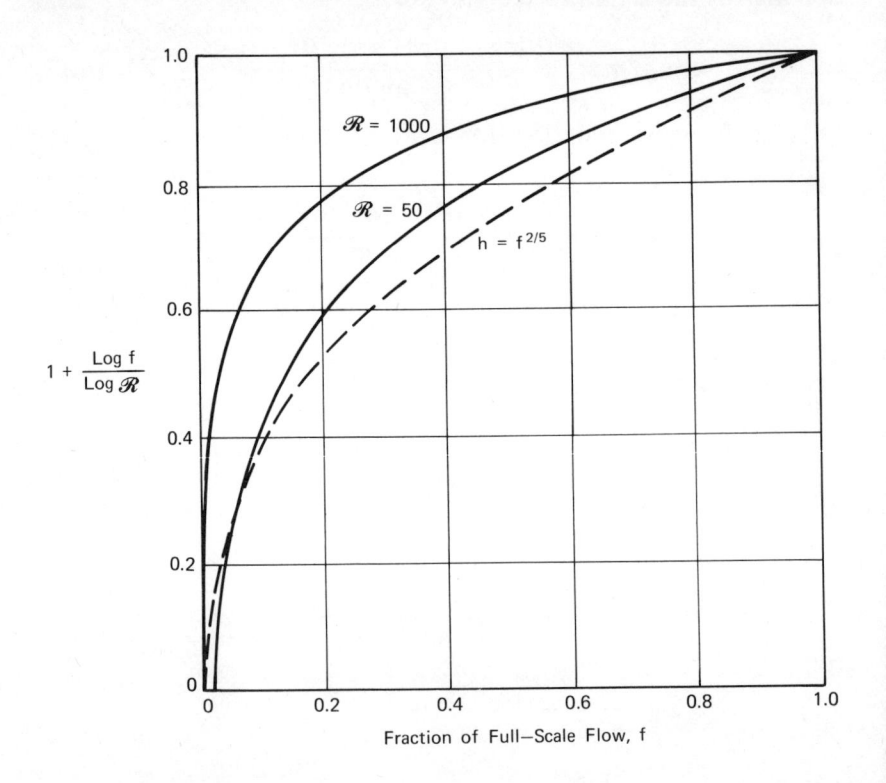

FIGURE 9.6. The functions of flow for 1000:1 and 50:1 range systems are compared to the characteristic of a Vee-notch weir.

percentage valve. If influent is flowing under level control from an upstream tank into the neutralizing vessel, the output of that controller (which is the position of the valve, if equipped with a positioner) represents the desired function of flow. If the influent valve's rangeability matches that of the reagent valve(s), no additional adjustment is required; however, if it has a lower rangeability, a gain equal to the ratio of the logarithms of the two rangeabilities must be applied. For example, a gain of 1.7 would have to be applied to match a 50:1 influent valve to a 1000:1 reagent system, as indicated by the difference between the two solid curves in Fig. 9.6.

In many, if not most wide-range systems, flow varies only 2 or 3:1 at most—a small fraction of the total rangeability of the plant. Furthermore, flow changes are usually not abrupt since the waste typically travels through open channels at least part of its way; in this way flow variations cause accompanying changes in level, which absorb any rapid fluctuations. For these cases, the function of flow may be fixed at 1.0, allowing feedback to correct for flow upsets.

The exception to this rule would be the plant fed by a lift pump (or pumps), periodically cycling on and off. In these cases, the flow signal could be made a stepwise function of the number of pumps in operation. Logic should be provided to shut off the flow of reagent altogether when pumps are off. The same signal may be used to transfer the feedback controller remotely to manual to avoid the reset windup described under the last heading in Chapter 8.

Returning to (9.15A) and (9.15B), separate out the flow functions:

$$m = 1 + \frac{\log f}{\log R} + \frac{\log(A/Bx_B) - \text{pH}}{\log R} \qquad (9.16\text{A})$$

$$m = 1 + \frac{\log f}{\log R} + \frac{\text{pH} - 14 + \log(B/Ax_A)}{\log R} \qquad (9.16\text{B})$$

Both equations fall into the general form:

$$m = b + \frac{100}{P}(r - c) \qquad (9.17)$$

where b is a bias term, $100/P$ is an adjustable gain, c is the pH input, and *r* is an adjustable reference point. The form of (9.17) has been intentionally selected to fit a proportional controller whose bias *b*, percent proportional band *P*, and set point *r* are all adjustable. In addition, the proportional controller may be transferred to manual if desired, to inhibit feedforward action for maintenance or test purposes.

The actual value of the proportional band, for a strong acid or base, is related to the span of the influent pH transmitter:

$$P = 100 \frac{\log R}{\text{pH span}} \quad (9.18)$$

For the example where $R = 1000$ and the range of the pH measurement is 0 to 10

$$P = 100 \frac{3}{10} = 30\%$$

In actual practice, *P* must be adjusted to provide a match for the normal or mean characteristic, but (9.18) serves as a firm starting point.

FEEDBACK TRIM

Feedforward control, though responsive, is never accurate enough to satisfy even broad limits of effluent pH. For example, strong acid at pH 2 must be neutralized to ±0.1% to meet limits of pH 5 to 9 and ±0.01% for pH 6 to 8. Carbonates and other buffers may moderate these sensitivities by a factor of 10 or even more, nonetheless present-day valves are probably no better than 5% accurate. (They may be more repeatable, but in a wide-range system overall accuracy is the critical factor, since reagent demand is continuously changing.)

The Effects of Weak Agents

But by far the greatest source of error is our inability to model the process exactly. If only strong acids and bases are present, or only a *single* weak acid or base, the relationship between pH and valve position is single-valued and linear. But even simple mixtures of strong agents with weak agents or their salts produce a multiplicity of curves, only two of which are shown in Fig. 9.2.

The response of the feedforward system compared to that of the process can be envisioned by following these curves. Let the feedforward system be calibrated for the strong-acid line, with an influent pH of 2. Should the influent contain a constant 10^{-3} N level of acetic acid, an increasing pH would cause the feedforward system to decrease the caustic flow too sharply, resulting in a low effluent pH. Observing this, the engineer might adjust his feedforward gain and set point to more nearly match the $x_C = 10^{-3}$ curve between pH 3 and 4. However, below pH 3 or above 5, the gain would be too low. Furthermore, a subsequent change in x_C will slide the curve up or down and also affect its slope.

Consider, however, the opposite situation wherein the concentration of acetic acid changes while that of the strong agents is fixed. Now a sizable change in reagent demand would accompany only a slight variation in pH, so that the feedforward gain would be too low, even if adjusted for strong acid alone.

If both the weak and strong agents change concentration by the same amount, as they would by dilution, a straight line is formed intermediate between the strong- and weak-acid lines. Although this condition is favorable for control, it cannot be expected in practice, since most waste-treatment plants receive influent from multiple sources. Variations in flow and/or composition of the individual sources—the most likely cause of load changes—will then affect the relative concentrations of weak and strong components.

Consequently, the engineer must adjust his feedforward settings to minimize the effect of the majority of the load changes, and rely on feedback to correct the balance. This is easier said than done, however. Some of the upsets are observable as variations in influent pH, while those caused by changes in weak-agent levels are not. In addition, the influent pH typically varies in a random pattern over which the engineer has no control. All he can do then is to adjust the feedforward settings to meet those load changes he observes, then by comparing the day-to-day or week-to-week pH records, evaluate their effectiveness.

Figure 9.7 compares influent and effluent pH records for a typical waste-treatment plant under feedforward control with feedback. Where the feedforward gain is too low for the process, undercorrection allows influent pH spikes to drive the effluent in

230 Feedforward and Adaptive Control

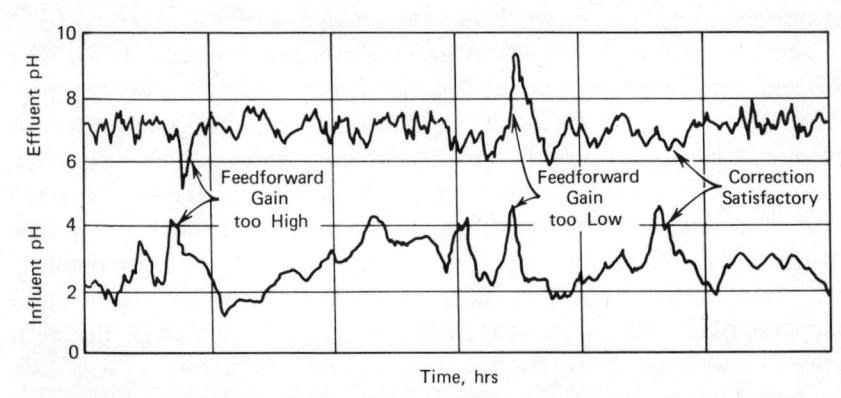

FIGURE 9.7. The proper adjustment of feedforward gain requires the analysis of hours or even days of operation.

the same direction; where feedforward gain is too high, effluent pH will go in the opposite direction from the influent. The engineer must review a day or more of operation, selecting his gain adjustment to distribute the peaks and valleys equally.

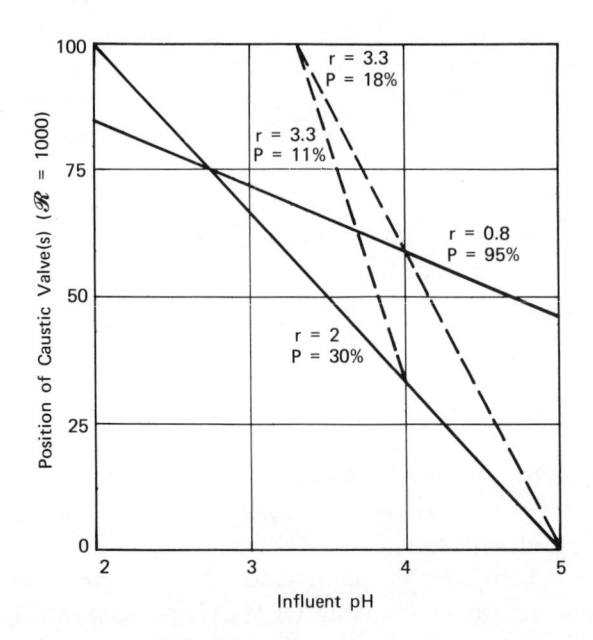

FIGURE 9.8. Shown are four possible combinations of feedforward parameters which match the curves of Fig. 9.2.

How to Introduce Feedback

Trying to match the variety of curves shown in Fig. 9.2 is a difficult task. Even using the linear approximations shown in Fig. 9.8, both slope and intercept must be adjusted independently. An imperfect calculation will cause a shift in effluent pH which the feedback controller can correct—but some decision must be made as to exactly how the feedback controller does it. Neither adjustment of the slope nor intercept alone is adequate to cover all the various combinations.

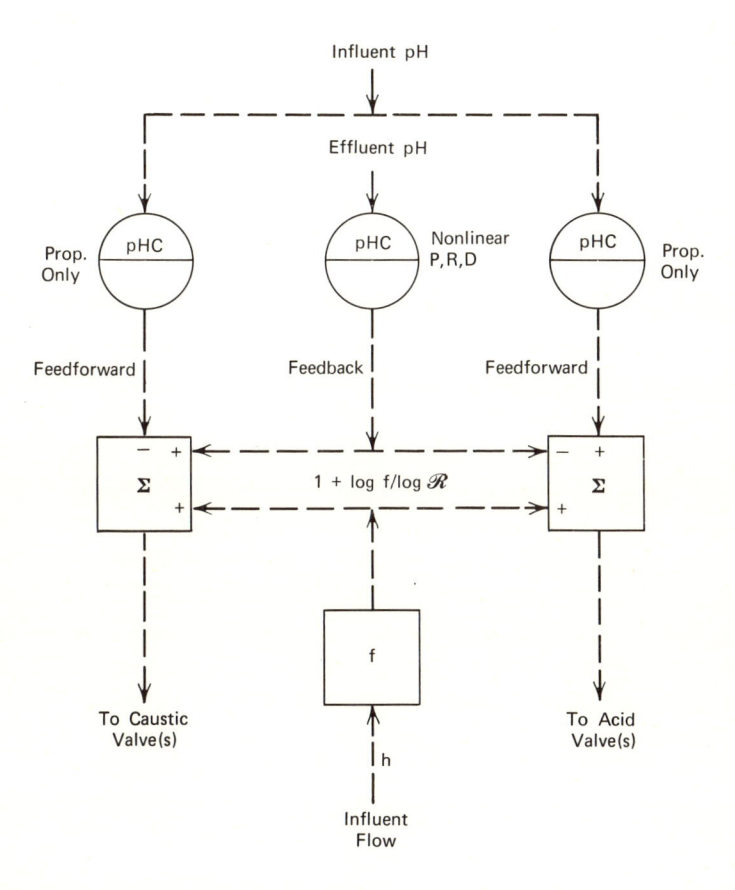

FIGURE 9.9. This configuration combines feedforward and feedback for the addition of both acid and basic reagents.

232 Feedforward and Adaptive Control

The best approach seems to be adjusting the feedforward gain manually and let the feedback controller move the intercept up or down as necessary to match the load. An arrangement combining the two loops is given in Fig. 9.9 for a system delivering both acid and basic reagents. The acid valve position m_A is

$$m_A = \left(1 + \frac{\log f}{\log R}\right) + \frac{100}{P_A}(r_A - \text{pH}) + m_F - 0.5 \quad (9.19A)$$

and the base valve position m_B is

$$m_B = \left(1 + \frac{\log f}{\log R}\right) - \frac{100}{P_B}(\text{pH} - r_B) - m_F + 0.5 \quad (9.19B)$$

Here m_F is the output of the feedback controller and 0.5 is a fixed bias in the summing device. The latter is intended to place the feedback term m_F at center scale (50%) for the normal case, allowing equal latitude to vary in either direction. The feedforward set points r_A and r_B should be adjusted so that the feedback controller output does not have to change greatly when the influent pH passes through neutrality. As noted earlier, if flow does not vary either widely or rapidly, the flow input may be left out of the summing units, with a bias of 1.0 inserted in its place.

FEEDFORWARD ADAPTATION

Earlier chapters have demonstrated how the control system should be matched to the needs and characteristics of the process for optimum performance. What, then, will happen when these characteristics change? If the control system is not able to change with them, either instability or poor response will result, depending on the direction of the change. But recently, automatic adaptation of controller parameters to match changing process characteristics has been successfully accomplished.

The application of adaptive control requires two features not normally available with conventional control instruments.

1. The controller settings must be capable of adjustment from a remote signal.

2. There must be some means of converting observations of the varying process characteristics into appropriate changes in control settings.

The second requirement may be fulfilled either with a feedforward or a feedback loop. With feedforward adaptation, the variable characteristics of the process are inferred from measurements made on its inputs. Feedback adaptation, however, is based on observations of control-system performance, as subject to variations in unmeasured or unmeasurable process properties. Feedforward adaptation is stable and responsive, but requires more information and is subject to inaccuracy or poor representation, as is any feedforward control system.

Variable Valve Gain

When *feedback* control systems for pH were described, it was pointed out that the control valves should be linear, except in the unusual case where a single weak acid or base was being neutralized. Yet *feedforward* control of pH over wide ranges requires that the valve or valves have *equal-percentage* characteristics. Feedback applied to one of these feedforward systems will then encounter variations in loop gain as the valve gain varies. Recall that the gain of an equal-percentage valve was proportional to its delivered flow as developed by (6.19). This means that the feedback loop gain in a feedforward system may vary as much as the rangeability of the system—50:1 with a single valve, and perhaps 1000:1 with two.

(This problem is not restricted to wide-range systems. In the narrow-range feedforward systems shown in Figs. 9.3 and 9.4, the output of the feedback controller is multiplied by influent flow, making its loop gain vary directly proportional to influent flow. The effect is the same, but much less severe than with the wide-range system, considering that flow may vary only over a 4:1 range.)

Adjusting the gain of a controller over a range of 1000:1 is neither easy nor practical. Fortunately, however, the characteristic of the nonlinear controller permits an appropriate adjustment to be implemented without great difficulty. To illustrate how this is carried out, compare the two titration curves given in Fig. 9.10. If a nonlinear controller were to be used to control both of the given influents, its deadband width should be varied as a function of influent pH.

However, the reason these two curves differ—while in fact they represent the same strong acid—is the selection of scales. The broken curve is simply an expansion of the other between the limits

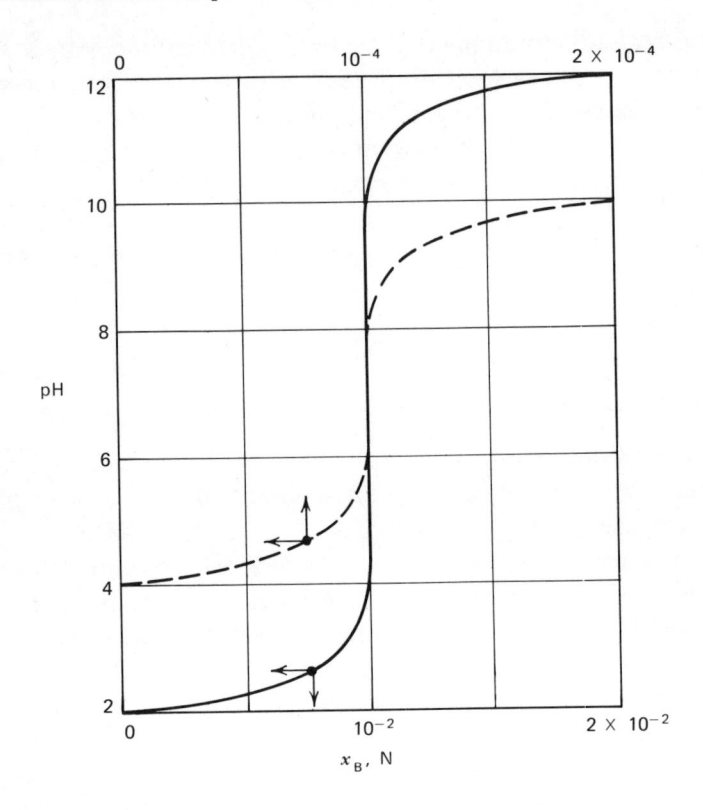

FIGURE 9.10. The deadband width should be related to influent pH.

of pH 4 and 10. This expansion is equivalent to neutralizing the more dilute (pH 4) influent with a valve 100 times smaller than that required for the more concentrated (pH 2) solution. Changing the size of a valve is therefore equivalent to changing the deadband width.

Actually, equal-percentage valves change their gain continuously with flow, as contrasted to linear valves whose gain only changes with size. Consequently, varying flow from an equal-percentage valve or valves could be represented by a variety of titration curves all representing different deadband widths.

The relationship between influent pH and deadband width is direct, and it effectively compensates for the problem of variable valve gain, as long as reagent flow is directly related to influent pH. Flow variations which affect reagent flow but not influent pH are

Feedforward Adaptation 235

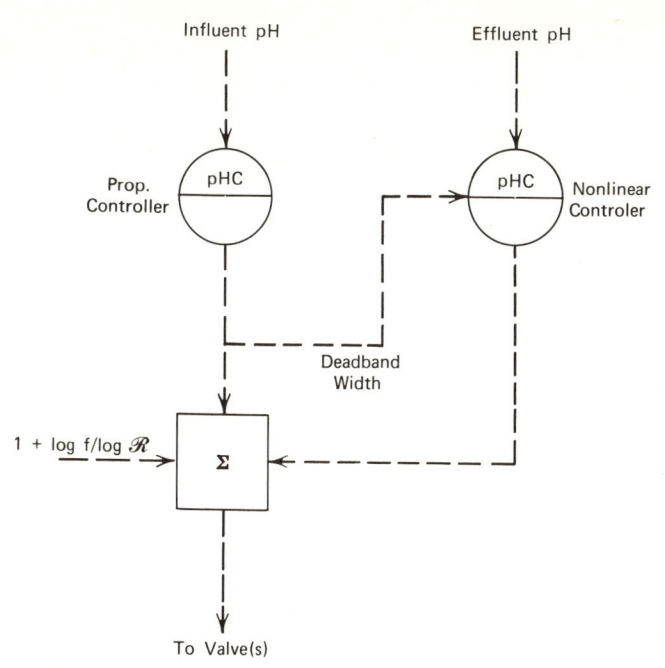

FIGURE 9.11. In this arrangement, influent pH is converted into equivalent valve position before being applied to the deadband width adjustment.

not thereby compensated. Changes in buffering are, however—increasing weak acid can require more caustic without measurably affecting influent pH, yet it will also moderate the titration curve in a compensating manner.

A feedforward system with feedforward adaptation of deadband width is shown in Fig. 9.11. Having first adjusted the feedforward controller, the effect of its output over the width of the deadband still must be set for feedback loop stability. A fully closed valve would call for zero deadband, while the width required for a fully open valve must be manually matched to the titration curve.

Variable Titration Curves

The optimum gains of both the feedforward and feedback controllers are affected by the process titration curve. If the curve of the influent were known at all times, both these gains could be adapted accordingly, and control would be close to perfect. Yet it is not actually necessary to know the entire curve to provide the required adaptation. The feedforward controller only requires a measurement of the influent normality to operate on—if this information is available, influent pH is unimportant and adaptation of the feedforward loop is actually unnecessary.

To adapt the feedback controller, only the neutral region of the titration curve must be known. A viable technique for achieving this adaptation in a feedforward manner involves operating a continuous automatic titrator with a linear controller adjusted to develop a limit cycle. The amplitude of the limit cycle could be used, after rectifying and filtering, to set the deadband width of the nonlinear controller.

The dynamic characteristics of such a feedforward adaptive loop are complex, since it contains an oscillating feedback loop. The period of the limit cycle must be short relative to that of the main feedback loop—this was pointed out earlier with regard to the response of the titrator in providing normality information for feedforward control. Filtering has a twofold effect: smoothing the deadband-width signal for stability, and modeling the time constant of the neutralization vessel. A first-order lag should be used, set equal to the time constant of the vessel, to refer influent information to the effluent, upon which the feedback controller is operating.

FEEDBACK ADAPTATION

A sizable percentage of waste-treatment facilities must accommodate influent streams with drastically varying properties. This is particularly true of plants with a multiplicity of products, each having its own combination of associated wastes. If some of these processes are operated batchwise, even in only one or two stages of production (such as washing or filtering), the influent titration curve will be affected.

The most common materials found in plant wastes are probably sulfuric acid and sodium hydroxide. Yet the weak acids and bases available to processors so outnumber the strong agents that the probability of their presence in any given waste stream is quite high. Appendix A lists the ionization constants of those weak agents most likely to be encountered—others exist which are considered so specialized that only the individual manufacturer need consider their properties.

Even in plants where only one or two products are manufactured, significant variations in influent properties could still appear. Periodic regeneration of ion-exchange beds, for example, can release copious doses of buffered salts into the waste-treating plant. And cleaning operations conducted at regular or irregular intervals can result in discharges of weak bases exhibiting a high degree of buffering. Phosphates, carbonates, silicates, and hypochlorites all have pK values near 7 and their presence can impose a massive gain reduction on an otherwise sensitive measurement.

These powerful buffers can adversely affect system performance if they appear at random in an otherwise unbuffered stream. Stability requires that the gain of the pH controller be set to a low value to accommodate the maximum gain of the process. The arrival of a weak base or acid then imposes a load increase and gain reduction simultaneously, driving the pH away from the set point with little reaction from the controller. A slow limit cycle could even develop as described in Fig. 8.7. In these situations, and there may be many across industry, adaptive control is essential.

Feedback would seem to be the only practical means of adaptation for variable buffering in view of questionable long-term reliability of automatic titrators at the present state of the art. Although other programmed types of adaptation[3] may be developed successfully for a given plant, feedback adaptation appears to be, by contrast, universally applicable.

Another Feedback Loop

Adaptation by feedback has the advantages and also the disadvantages shared by other feedback control loops. By observing the actual performance of the control system, adjustments can be made to modify it to some more desirable state. The goal is not necessarily attainable at any given time, however, just as straight-line control may only occasionally be achieved by the feedback pH controller, due to upsets, nonidealities, and so on.

Furthermore, the adaptive loop is just as prone to instability as the pH control loop, with a period that may be four to ten times as long. Finally, adaptation is made after the fact; it represents condi-

238 Feedforward and Adaptive Control

tions which have preceded, and which may no longer be in effect. Consequently, only steady-state changes in gain can be adapted—variations occurring pulsewise cannot hope to be accommodated by feedback.

For the same reasons given under feedforward adaptation, the deadband width is the adjustment also selected for feedback adaptation. Figure 9.12 shows how the adaptive controller would be configured with a nonlinear pH controller. Although at this writing, vitrually every application of feedback adaptation to process control has been by means of a digital computer, an analog system will be described here. The fact that adaptive control has been successfully achieved with analog instruments[4] is of singular importance, in that its principal area of application, that is, waste-treatment plants, can rarely justify a digital computer for control.

Assimilating Performance

Desirable response for a pH loop may differ considerably from that expected from other processes. This is particularly true with regard to waste treatment. For example, any pH between 6 and 9 may meet local regulations and therefore be acceptable. Limit-cycling between the established limits may also be acceptable, *but only if a single reagent is used.* Dual-reagent systems have been observed

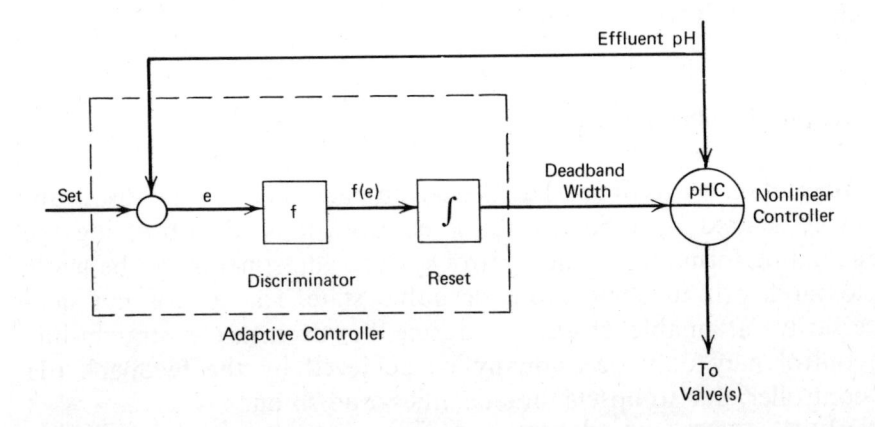

FIGURE 9.12. The adaptive controller observes the stability of the pH loop and adjusts the nonlinear controller accordingly.

limit-cycling well within established limits, but at the same time consuming reagent even with no load on the plant. Needless to say, excessive reagent usage is costly both from a supply and a disposal standpoint—everything that enters the plant must leave it, either in solution or as a sludge. (It is possible to conserve reagent and tolerate cycling in a dual-reagent system by inhibiting the basic reagent valve when the influent pH is above 7 and the acid valve when the influent pH is below 7. Operation of such a system is not as simple as it would seem, however. Upsets often cause the controller to add too much reagent; when this occurs with the other reagent inhibited, recovery could be unacceptably slow. In addition, controller windup would develop during these intervals, and unless prevented by additional logic, would cause a severe upset when the influent pH crossed 7. Experience with complex logic systems to perform these functions has not been entirely satisfactory.)

For a dual-reagent system, then, cycling is to be avoided as well as any pH outside of established limits. But allowing an offset near or at a limit is poor practice since little margin is thereby allowed to absorb an upset which will surely come. The adaptive controller should therefore have an offset criterion, although not necessarily linear. Offset approaching a limit should induce substantial action; a small offset need not induce any.

Increasing loop gain either through the process or by narrowing the deadband will eventually develop a limit cycle at the natural period of the loop. *Reducing* loop gain by either manipulation will cause an increasing—but not sustained—offset. Reset action eventually will bring the measurement back toward the set point, and in fact, may develop a cycle of much longer period. This phenomenon was described as conditional stability in Chapter 8, during the discussion of the nonlinear controller. Because the period of the "low-gain" limit cycle may be as great as ten times the natural period of the more familiar "high-gain" cycle, the two are readily distinguishable. The slower cycle may be eliminated by closing the deadband width, increasing the deadband gain, or reducing the proportional band, all of which increase the loop gain to smaller amplitude signals.

Adaptive action then must discriminate between oscillations at or near the natural period, and those which are much longer. Offset or sluggishness, although not periodic, will naturally be included in the latter group. The characteristic desired for the discriminator is given

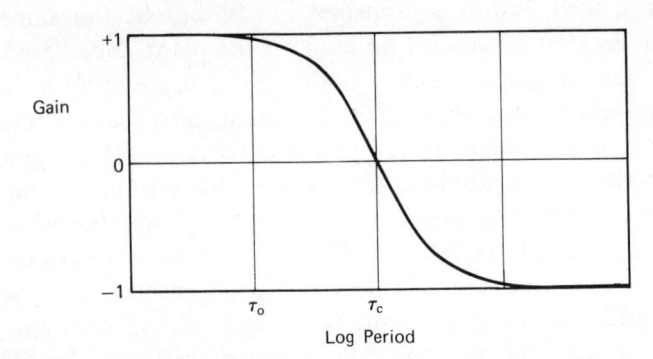

FIGURE 9.13. The discriminator features a gain reversal at the crossover period, τ_c.

in Fig. 9.13. Its crossover period τ_c must be adjustable to allow placing the natural period τ_o in the region of positive gain. The discriminator will only be satisfied with a deviation of zero or an oscillation precisely at the crossover period.

Adaptive-Loop Stability

Proportional action is not of any particular value in feedback adaptation. Reset must be used so that the deadband width can be sought which best meets the performance criterion, whatever its value. But the width adjustment must be made gradually, otherwise it will interfere with the normal operation of the pH control loop, tending to distort its response and possibly destabilize or desensitize the pH controller. The dominance of the reset mode will force the period of the adaptive loop to be a multiple of the pH loop natural period.

Reset windup must be guarded against in the adaptive loop—the controller should be automatically transferred to manual whenever the pH controller is in manual or is otherwise incapable of regulation. If this is not done, the adaptive controller will tend to reduce the deadband to zero (or to whatever low limit is placed on it), causing an oscillation to develop when automatic control is restored.

If the reset time of the adaptive controller is set too low, the adaptive loop will become unstable, causing a periodic pattern of cycling in the pH record. This characteristic, called "burst cycling,"

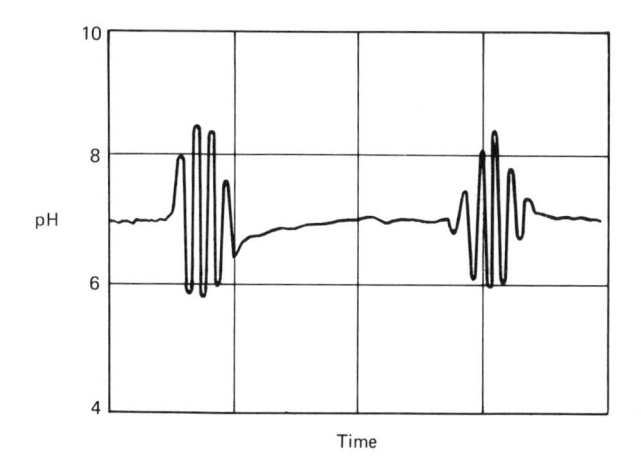

FIGURE 9.14. Burst cycling is characteristic of an unstable adaptive feedback loop.

is shown in Fig. 9.14. A developing cycle can be quenched by the adaptive controller but any overcorrection will result in sluggishness, causing the deadband to be again contracted.

Even with proper reset adjustment, however, the extreme sensitivity of pH to reagent flow in the eutral region may preclude the complete elimination of offset. Consequently, to avoid burst cycling requires some sort of deadband to be applied to the adaptive controller. A deadband may be superimposed on the deviation, which for single-reagent systems, may be sufficient. However, a dual-reagent system, while accepting an offset, cannot tolerate even a small limit cycle. The optimum arrangement for all systems has been found to be an adjustable gain applied only to the low-frequency portion of the spectrum. This confirms the dual criterion of defined limits for offset with no tolerance for oscillation. This asymmetry is necessary to make the adaptive loop stable.

SUMMARY

Feedback adaptation is by no means a panacea. In automating one adjustment (deadband width), we have added three more (crossover period, low-frequency gain, and adaptive reset). Each step we take in extending our technology requires more support, more components, more attention, and tends to be less reliable. Every effort should be made to keep our processes as simple as possible consistent with their objectives. A well-stirred tank which is easily controlled will always give more satisfactory results than a plug-flow process fitted with a complex system of instruments.

The following quotation by Elting Morison may be applied to all phases of our technology:

> In achieving control over our resistant natural environment, we find we have produced in the means themselves, an artificial environment of such complexity that we cannot control it.

The implication here is that we are actually losing ground in our efforts to improve the quality of life. However, the outlook cannot be so bleak: Two steps backward for every three forward seems a more representative measurement of our technological progress.

REFERENCES

1. F. G. Shinskey, *Process-Control Systems*, McGraw-Hill Book Company, New York, 1967, pp. 204–228.
2. L. K. Spink, *Principles and Practices of Flowmeter Engineering*, 9th ed., The Foxboro Company, Foxboro, Massachusetts, 1967.
3. V. L. Trevathan, pH Control in Waste Streams, Instrument Society of America Paper No. 72-725.
4. F. G. Shinskey, A Self-Adjusting System for Effluent pH Control, presented at the 1973 Instrument Society of America Joint Spring Conference, St. Louis, Missouri, April 24–26, 1973.

APPENDIX A
IONIZATION CONSTANTS OF ACIDS AND BASES

Acid	Equilibrium	pK_A	Reference
Acetamide	$CH_3CONH_2 \rightleftarrows CH_3CONH^- + H^+$	0.63	5
Acetic acid	$CH_3COOH \rightleftarrows CH_3COO^- + H^+$	4.75	1
Aluminum ion	$Al^{3+} + H_2O \rightleftarrows AlOH^{2+} + H^+$	5.1	2
Aluminum hydroxide	$Al(OH)_3 \rightleftarrows AlO_2^- + H_2O + H^+$	12.4	3
Ammonium ion	$NH_4^+ \rightleftarrows NH_3 + H^+$	9.25	1
Anilinium ion	$C_6H_5NH_3^+ \rightleftarrows C_6H_5NH_2 + H^+$	4.7	4
Boric acid	$H_3BO_3 \rightleftarrows H_2BO_3^- + H^+$	9.1	4
	$H_2BO_3^- \rightleftarrows HBO_3^{2-} + H^+$	12.7	5
	$HBO_3^{2-} \rightleftarrows BO_3^{3-} + H^+$	13.8	5
Benzene sulfonic acid	$C_6H_5SO_3H \rightleftarrows C_6H_5SO_3^- + H^+$	0	2
Benzoic acid	$C_6H_5COOH \rightleftarrows C_6H_5COO^- + H^+$	4.1	4
Carbon dioxide	$CO_2 + H_2O \rightleftarrows HCO_3^- + H^+$	6.35	1
	$HCO_3^- \rightleftarrows CO_3^{2-} + H^+$	10.25	1
Chromic acid	$H_2CrO_4 \rightleftarrows HCrO_4^- + H^+$	0.7	4
	$HCrO_4^- \rightleftarrows CrO_4^{2-} + H^+$	6.2	4
Chromium ion	$Cr^{3+} + H_2O \rightleftarrows CrOH^{2+} + H^+$	4.2	2
	$CrOH^{2+} + H_2O \rightleftarrows Cr(OH)_2^+ + H^+$	6.2	2
Chromium hydroxide	$Cr(OH)_3 \rightleftarrows CrO_2^- + H_2O + H^+$	16	3
Chloracetic acid	$CH_2ClCOOH \rightleftarrows CH_2ClCOO^- + H^+$	2.7	4
Citric acid	$C_6H_8O_7 \rightleftarrows C_6H_7O_7^- + H^+$	3.0	4
	$C_6H_7O_7^- \rightleftarrows C_6H_6O_7^{2-} + H^+$	4.4	4
	$C_6H_6O_7^{2-} \rightleftarrows C_6H_5O_7^{3-} + H^+$	6.1	4
	$C_6H_5O_7^{3-} \rightleftarrows C_6H_4O_7^{4-} + H^+$	16	1

243

Acid	Equilibrium	pK_A	Reference
Cupric ion	$Cu^{2+} + H_2O \rightleftharpoons CuOH^+ + H^+$	6.8	2
Cyanic acid	$HCNO \rightleftharpoons CNO^- + H^+$	3.6	4
Dichloracetic acid	$CHCl_2COOH \rightleftharpoons CHCl_2COO^- + H^+$	1.1	4
Ferric ion	$Fe^{3+} + H_2O \rightleftharpoons FeOH^{2+} + H^+$	2.5	2
	$FeOH^{2+} + H_2O \rightleftharpoons Fe(OH)_2^+ + H^+$	4.7	2
Ferrous ion	$Fe^{2+} + H_2O \rightleftharpoons FeOH^+ + H^+$	6.8	2
Formic acid	$HCOOH \rightleftharpoons HCOO^- + H^+$	3.65	4
Gluconic acid	$C_6H_{12}O_7 \rightleftharpoons C_6H_{11}O_7^- + H^+$	4.7	2
Hydrogen cyanide	$HCN \rightleftharpoons CN^- + H^+$	9.2	4
Hydrogen fluoride	$HF \rightleftharpoons F^- + H^+$	3.17	1
	$HF + F^- \rightleftharpoons HF_2^-$	0.6	4
Hydrogen sulfide	$H_2S \rightleftharpoons HS^- + H^+$	7.0	1
	$HS^- \rightleftharpoons S^{2-} + H^+$	12.9	1
Hydroquinone	$C_6H_6O_2 \rightleftharpoons C_6H_5O_2^- + H^+$	10.4	5
Hydroxylammonium ion	$NH_3OH^+ \rightleftharpoons NH_2OH + H^+$	6.2	4
Hypochlorous acid	$HClO \rightleftharpoons ClO^- + H^+$	7.5	1
Iodic acid	$HIO_3 \rightleftharpoons IO_3^- + H^+$	0.79	1
Nitrous acid	$HNO_2 \rightleftharpoons NO_2^- + H^+$	3.2	4
Oxalic acid	$H_2C_2O_4 \rightleftharpoons HC_2O_4^- + H^+$	1.1	4
	$HC_2O_4^- \rightleftharpoons C_2O_4^{2-} + H^+$	4.0	4
Phenol	$C_6H_5OH \rightleftharpoons C_6H_5O^- + H^+$	9.8	4
Phosphoric acid	$H_3PO_4 \rightleftharpoons H_2PO_4^- + H^+$	2.23	1
	$H_2PO_4^- \rightleftharpoons HPO_4^{2-} + H^+$	7.21	1
	$HPO_4^{2-} \rightleftharpoons PO_4^{3-} + H^+$	12.32	1
Phosphorous acid	$H_3PO_3 \rightleftharpoons H_2PO_3^- + H^+$	2.0	4
	$H_2PO_3^- \rightleftharpoons HPO_3^{2-} + H^+$	6.4	4
Phthalic acid	$C_8H_6O_4 \rightleftharpoons C_8H_5O_4^- + H^+$	2.8	4
	$C_8H_5O_4^- \rightleftharpoons C_8H_4O_4^{2-} + H^+$	5.1	4
Propionic acid	$C_2H_5COOH \rightleftharpoons C_2H_5COO^- + H^+$	4.9	3
Silicic acid	$Si(OH)_4 \rightleftharpoons SiO(OH)_3^- + H^+$	9.6	4
	$SiO(OH)_3^- \rightleftharpoons SiO_2(OH)_2^{2-} + H^+$	12.7	4
Sulfamic acid	$NH_2SO_3H \rightleftharpoons NH_2SO_3^- + H^+$	1.0	4
Sulfanilic acid	$NH_2C_6H_5SO_3H \rightleftharpoons NH_2C_6H_5SO_3^- + H^+$	3.2	2
Sulfuric acid	$H_2SO_4 \rightleftharpoons HSO_4^- + H^+$	-3	4
	$HSO_4^- \rightleftharpoons SO_4^{2-} + H^+$	1.99	1
Sulfurous acid	$H_2SO_3 \rightleftharpoons HSO_3^- + H^+$	1.8	4
	$HSO_3^- \rightleftharpoons SO_3^{2-} + H^+$	6.8	4
Tartaric acid	$C_4H_6O_6 \rightleftharpoons C_4H_5O_6^- + H^+$	2.9	4
	$C_4H_5O_6^- \rightleftharpoons C_4H_4O_6^{2-} + H^+$	4.1	4
Thiosulfuric acid	$H_2S_2O_3 \rightleftharpoons HS_2O_3^- + H^+$	0.6	1
	$HS_2O_3^- \rightleftharpoons S_2O_3^{2-} + H^+$	1.7	1
Trichloracetic acid	$CCl_3COOH \rightleftharpoons CCl_3COO^- + H^+$	0.5	4
Uric acid	$C_5H_4N_4O_3 \rightleftharpoons C_5H_3N_4O_3^- + H^+$	3.9	5
Xanthine	$C_5H_4N_4O_2 \rightleftharpoons C_5H_3N_4O_2^- + H^+$	9.9	5

Ionization Constants of Acids and Bases

Base	Equilibrium	pK_B	Reference
Acetamide	$CH_3CONH_2 + H_2O \rightleftharpoons CH_3CONH_3^+ + OH^-$	13.4	5
Aluminate ion	$AlO_2^- + 2H_2O \rightleftharpoons Al(OH)_3 + OH^-$	1.6	3
Aluminum hydroxide	$AlOH^{2+} \rightleftharpoons Al^{3+} + OH^-$	8.9	2
Ammonia	$NH_3 + H_2O \rightleftharpoons NH_4^+ + OH^-$	4.75	1
Aniline	$C_6H_5NH_2 + H_2O \rightleftharpoons C_6H_5NH_3^+ + OH^-$	9.38	1
Barium hydroxide	$Ba(OH)_2 \rightleftharpoons BaOH^+ + OH^-$	0.7	2
Calcium hydroxide	$Ca(OH)_2 \rightleftharpoons CaOH^+ + OH^-$	1.40	5
	$CaOH^+ \rightleftharpoons Ca^{2+} + OH^-$	2.43	5
Chromium hydroxide	$Cr(OH)_2^+ \rightleftharpoons CrOH^{2+} + OH^-$	7.8	2
	$CrOH^+ \rightleftharpoons Cr^{3+} + OH^-$	9.8	2
Cupric hydroxide	$CuOH^+ \rightleftharpoons Cu^{2+} + OH^-$	7.2	2
Ethylamine	$C_2H_5NH_2 + H_2O \rightleftharpoons C_2H_5NH_3^+ + OH^-$	3.3	3
Ferric hydroxide	$Fe(OH)^+ \rightleftharpoons FeOH^{2+} + OH^-$	9.3	2
	$FeOH^{2+} \rightleftharpoons Fe^{3+} + OH^-$	11.5	2
Ferrous hydroxide	$FeOH^+ \rightleftharpoons Fe^{2+} + OH^-$	7.2	2
Hydrazine	$NH_2NH_2 + H_2O \rightleftharpoons NH_2NH_3^+ + OH^-$	5.5	3
Hydroxylamine	$NH_2OH + H_2O \rightleftharpoons NH_3OH^+ + OH^-$	7.97	5
Magnesium hydroxide	$Mg(OH)_2 \rightleftharpoons MgOH^+ + OH^-$	2.6	2
Methylamine	$CH_3NH_2 + H_2O \rightleftharpoons CH_3NH_3^+ + OH^-$	3.28	1
Urea	$NH_2CONH_2 + H_2O \rightleftharpoons NH_2CONH_3^+ + OH^-$	13.9	5
Zinc hydroxide	$Zn(OH)_2 \rightleftharpoons ZnOH^+ + OH^-$	3.02	5

REFERENCES

1. J. N. Butler, *Ionic Equilibrium—A Mathematical Approach*, Addison-Wesley, Reading, Massachusetts, 1964.
2. *Stability Constants of Metal-Ion Complexes*, The Chemical Society, London, 1964.
3. T. R. Hogness and W. C. Johnson, *Qualitative Analysis and Chemical Equilibrium*, 3rd ed., Henry Holt and Company, New York, 1947.
4. A. Ringbom, *Complexation in Analytical Chemisty*, Interscience Publishers, New York, 1963.
5. R. C. Weast, *Handbook of Chemistry and Physics*, The Chemical Rubber Company, Cleveland, Ohio, 1970.

APPENDIX B
SOLUBILITY PRODUCT CONSTANTS

Substance	Equilibrium	pK_{sp}	Reference
Aluminum hydroxide	$Al(OH)_3 \rightleftarrows Al^{3+} + 3OH^-$	33	3
Barium carbonate	$BaCO_3 \rightleftarrows Ba^{2+} + CO_3^{2-}$	8.1	3
Barium chromate	$BaCrO_4 \rightleftarrows Ba^{2+} + CrO_4^{2-}$	9.6	3
Barium fluoride	$BaF_2 \rightleftarrows Ba^{2+} + 2F^-$	5.8	3
Barium sulfate	$BaSO_4 \rightleftarrows Ba^{2+} + SO_4^{2-}$	10	3
Cadmium hydroxide	$Cd(OH)_2 \rightleftarrows Cd^{2+} + 2OH^-$	13.9	3
Cadmium sulfide	$CdS \rightleftarrows Cd^{2+} + S^{2-}$	28	3
Calcium carbonate	$CaCO_3 \rightleftarrows Ca^{2+} + CO_3^{2-}$	8.1	3
Calcium fluoride	$CaF_2 \rightleftarrows Ca^{2+} + 2F^-$	10.4	3
Calcium phosphate	$Ca_3(PO_4)_2 \rightleftarrows 3Ca^{2+} + 2PO_4^{3-}$	26	1
	$CaHPO_4 \rightleftarrows Ca^{2+} + HPO_4^{2-}$	8.4	1
Calcium sulfate	$CaSO_4 \rightleftarrows Ca^{2+} + SO_4^{2-}$	4.2	3
Calcium sulfite	$CaSO_3 \rightleftarrows Ca^{2+} + SO_3^{2-}$	5.7	4
Chromium hydroxide	$Cr(OH)_3 \rightleftarrows Cr^{3+} + 3OH^-$	30	3
Cupric hydroxide	$Cu(OH)_2 \rightleftarrows Cu^{2+} + 2OH^-$	19.2	3
Cupric sulfide	$CuS \rightleftarrows Cu^{2+} + S^{2-}$	37.5	3
Ferric hydroxide	$Fe(OH)_3 \rightleftarrows Fe^{3+} + 3OH^-$	35.8	3
Ferric ferrocyanide	$Fe_4[Fe(CN)_6]_3 \rightleftarrows 4Fe^{3+} + 3[Fe(CN)_6]^{4-}$	40	4
Ferrous carbonate	$FeCO_3 \rightleftarrows Fe^{2+} + CO_3^{2-}$	14.7	4
Ferrous hydroxide	$Fe(OH)_2 \rightleftarrows Fe^{2+} + 2OH^-$	13.7	3
Ferrous sulfide	$FeS \rightleftarrows Fe^{2+} + S^{2-}$	18.4	3
Lead carbonate	$PbCO_3 \rightleftarrows Pb^{2+} + CO_3^{2-}$	13.4	3
Lead chromate	$PbCrO_4 \rightleftarrows Pb^{2+} + CrO_4^{2-}$	13.7	3
Lead hydroxide	$Pb(OH)_2 \rightleftarrows Pb^{2+} + 2OH^-$	15.6	3
Lead sulfate	$PbSO_4 \rightleftarrows Pb^{2+} + SO_4^{2-}$	8	3
Lead sulfide	$PbS \rightleftarrows Pb^{2+} + S^{2-}$	29.2	3
Magnesium carbonate	$MgCO_3 \rightleftarrows Mg^{2+} + CO_3^{2-}$	4.4	3

Solubility Product Constants

Substance	Equilibrium	pK_{sp}	Reference
Magnesium fluoride	$MgF_2 \rightleftarrows Mg^{2+} + 2F^-$	8.2	3
Magnesium hydroxide	$Mg(OH)_2 \rightleftarrows Mg^{2+} + 2OH^-$	10.8	3
Mercurous chloride	$Hg_2Cl_2 \rightleftarrows Hg_2^{2+} + 2Cl^-$	17.8	3
Mercuric hydroxide	$HgO + H_2O \rightleftarrows Hg^{2+} + 2OH^-$	21.7	3
Mercuric sulfate	$Hg_2SO_4 \rightleftarrows 2Hg^+ + SO_4^{2-}$	6.17	1
Mercuric sulfide	$Hg_2S \rightleftarrows 2Hg^+ + S^{2-}$	50	3
Nickel hydroxide	$Ni(OH)_2 \rightleftarrows Ni^{2+} + 2OH^-$	13.7	3
Nickel sulfide	$NiS \rightleftarrows Ni^{2+} + S^{2-}$	23.9	3
Silver bromide	$AgBr \rightleftarrows Ag^+ + Br^-$	12.1	3
Silver carbonate	$Ag_2CO_3 \rightleftarrows 2Ag^+ + CO_3^{2-}$	11.2	3
Silver chloride	$AgCl \rightleftarrows Ag^+ + Cl^-$	9.8	3
Silver chromate	$Ag_2CrO_4 \rightleftarrows 2Ag^+ + CrO_4^{2-}$	11	3
Silver cyanide	$AgCN \rightleftarrows Ag^+ + CN^-$	11.7	3
Silver dichromate	$Ag_2Cr_2O_7 \rightleftarrows 2Ag^+ + Cr_2O_7^{2-}$	6.7	3
Silver hydroxide	$\tfrac{1}{2}Ag_2O + \tfrac{1}{2}H_2O \rightleftarrows Ag^+ + OH^-$	7.8	3
Silver iodide	$AgI \rightleftarrows Ag^+ + I^-$	15.8	3
Silver sulfate	$Ag_2SO_4 \rightleftarrows 2Ag^+ + SO_4^{2-}$	4.8	1
Silver sulfide	$Ag_2S \rightleftarrows 2Ag^+ + S^{2-}$	51.4	3
Silver thiocyanate	$AgCNS \rightleftarrows Ag^+ + CNS^-$	11.9	3
Zinc carbonate	$ZnCO_3 \rightleftarrows Zn^{2+} + CO_3^{2-}$	10	4
Zinc hydroxide	$Zn(OH)_2 \rightleftarrows Zn^{2+} + 2OH^-$	16.3	3
Zinc sulfide	$ZnS \rightleftarrows Zn^{2+} + S^{2-}$	22.9	3

REFERENCES

1. J. N. Butler, *Ionic Equilibrium—A Mathematical Approach*, Addison-Wesley, Reading, Massachusetts, 1964.
2. *Stability Constants of Metal-Ion Complexes*, The Chemical Society, London, 1964.
3. T. R. Hogness and W. C. Johnson, *Qualitative Analysis and Chemical Equilibrium*, 3rd ed., Henry Holt and Company, New York, 1947.
4. A. Ringbom, *Complexation in Analytical Chemisty*, Interscience Publishers, New York, 1963.

APPENDIX C
STANDARD REDUCTION POTENTIALS[1]

Reduction Equation	Potential (mV)
$Ag^+ + e \to Ag$	+799.6
$AgCl + e \to Ag + Cl^-$	+222.3
Silver-silver chloride electrode 0.1 M KCl	+293.8
Silver-silver chloride electrode 1 M KCl	+235.8
Silver-silver chloride electrode 4 M KCl	+199.8
Silver-silver chloride electrode saturated KCl	+198.5
$Ag_2S + 2e \to 2Ag + S^{2-}$	−705.1
$Al^{3+} + 3e \to Al$	−1706
$H_2AlO_3 + H_2O + 3e \to Al + 4OH^-$	−2350
$As_2O_3 + 6H^+ + 6e \to 2As + 3H_2O$	+234
$Au^+ + e \to Au$	+1680
$Ba^{2+} + 2e \to Ba$	−2900
$Be^{2+} + 2e \to Be$	−1700
$Br_2 + 2e \to 2Br^-$	+1065
$HBrO + H^+ + 2e \to Br^- + H_2O$	+1330
$Ca^{2+} + 2e \to Ca$	−2760
Calomel electrode 1 N KCl	+280.7
Calomel electrode 0.1 N KCl	+333.7
Calomel electrode saturated KCl	+241.5
$Cd^{2+} + 2e \to Cd$	−402.6
$Cl_2(g) + 2e \to 2Cl^-$	+1358.3
$HOCl + H^+ + e \to Cl_2/2 + H_2O$	+1630
$HOCl + H^+ + e \to Cl^- + H_2O$	+1490

Reduction Equation	Potential (mV)
$ClO^- + H_2O + 2e \rightarrow Cl^- + 2OH^-$	+900
$ClO_2 + e \rightarrow ClO_2^-$	+1150
$ClO_2^- + 2H_2O + 4e \rightarrow Cl^- + 4OH^-$	+760
$ClO_3^- + 6H^+ + 6e \rightarrow Cl^- + 3H_2O$	+1450
$ClO_4^- + 8H^+ + 8e \rightarrow Cl^- + 4H_2O$	+1370
$(CN)_2 + 2H^+ + 2e \rightarrow 2HCN$	+370
$2HCNO + 2H^+ + 2e \rightarrow (CN)_2 + 2H_2O$	+330
$Co^{2+} + 2e \rightarrow Co$	−280
$CO_2 + 2H^+ + 2e \rightarrow HCOOH$	−200
$2CO_2 + 2H^+ + 2e \rightarrow H_2C_2O_4$	−490
$Cr^{3+} + 3e^- \rightarrow Cr$	−740
$Cr_2O_7^{2-} + 14H^+ + 6e \rightarrow 2Cr^{3+} + 7H_2O$	+1330
$HCrO_4^- + 7H^+ + 3e \rightarrow Cr^{3+} + 4H_2O$	+1195
$Cu^+ + e \rightarrow Cu$	+522
$Cu^{2+} + e \rightarrow Cu^+$	+158
$Cu^{2+} + 2e \rightarrow Cu$	+340.2
$\frac{1}{2} F_2 + e \rightarrow F^-$	+2850
$\frac{1}{2} F_2 + H^+ + e \rightarrow HF$	+3030
$Fe^{2+} + 2e \rightarrow Fe$	−409
$Fe^{3+} 3e \rightarrow Fe$	−36
$Fe^{3+} + e \rightarrow Fe^{2+}$	+770
$Fe(CN)_6^{3-} + e \rightarrow Fe(CN)_6^{4-}$	+500
$2H^+ + 2e \rightarrow H_2$	0.0
$2H_2O + 2e \rightarrow H_2 + 2OH^-$	−827.7
$H_2O_2 + 2H^+ + 2e \rightarrow 2H_2O$	+1776
$Hg^{2+} + 2e \rightarrow Hg$	+851
$2Hg^{2+} + 2e \rightarrow Hg_2^{2+}$	+905
$Hg_2Cl_2 + 2e \rightarrow 2Hg + 2Cl^-$	+268.2
$I_2 + 2e \rightarrow 2I^-$	+535
$HIO + H^+ + 2e \rightarrow I^- + H_2O$	+990
$IO_3^- + 6H^+ + 6e \rightarrow I^- + 3H_2O$	+1085
$K^+ + e \rightarrow K$	−2924
$Li^+ + e \rightarrow Li$	−3045
$Mg^{2+} + 2e \rightarrow Mg$	−2375
$Mn^{2+} + 2e \rightarrow Mn$	−1029
$MnO_2 + 4H^+ + 2e^- \rightarrow Mn^{2+} + 2H_2O$	+1208
$MnO_4^- + 4H^+ + 3e^- \rightarrow MnO_2 + 2H_2O$	+1679
$MnO_4^- + 8H^+ + 5e^- \rightarrow Mn^{2+} + 4H_2O$	+1491
$N_2 + 2H_2O + 4H^+ + 2e \rightarrow 2NH_3OH^+$	−1870
$3N_2 + 2H^+ + 2e \rightarrow 2NH_3$	−3100
$N_2O + 2H^+ + 2e \rightarrow N_2 + H_2O$	+1770
$NO_3^- + H_2O + 2e \rightarrow NO_2^- + 2OH^-$	+10
$Na^+ + e \rightarrow Na$	−2711
$Ni^{2+} + 2e \rightarrow Ni$	−230

Reduction Equation	Potential (mV)
$O_2 + 2H_2O + 4e \rightarrow 4OH^-$	+401
$O_3 + 2H^+ + 2e \rightarrow O_2 + H_2O$	+2070
$Pb^{2+} + 2e \rightarrow Pb$	−126.3
$Pd^{2+} + 2e \rightarrow Pd$	+830
$H_3PO_3 + 2H^+ + 2e \rightarrow H_3PO_2 + H_2O$	−500
$H_3PO_4 + 2H^+ + 2e \rightarrow H_3PO_3 + H_2O$	−276
$PtCl_4^{2-} + 2e \rightarrow Pt + 4Cl^-$	+730
Quinhydrone electrode $[H^+] = 1$	+699.5
$S + 2e \rightarrow S^{2-}$	−508
$S + 2H^+ + 2e \rightarrow H_2S(aq)$	+141
$Sb_2O_3 + 6H^+ + 6e \rightarrow 2Sb + 3H_2O$	+144.5
$SiF_6^{2-} + 4e \rightarrow Si + 6F^-$	−1200
$SiO_2 + 4H^+ + 4e \rightarrow Si + 2H_2O$	−840
$Sn^{2+} + 2e \rightarrow Sn$	−136.4
$Sn^{4+} + 2e \rightarrow Sn^{2+}$	+150
$2SO_3^{2-} + 3H_2O + 4e \rightarrow S_2O_3^{2-} + 6OH$	−580
$SO_4^{2-} + 4H^+ + 2e \rightarrow H_2SO_3 + H_2O$	+200
$SO_4^{2-} + 2H^+ + 2e \rightarrow SO_3^{2-} + H_2O$	+170
$Ti^{2+} + 2e \rightarrow Ti$	−1630
$Th^{4+} + 4e \rightarrow Th$	−1900
Thalamid electrode	−572
$TiO_2 + 4H^+ + 4e \rightarrow Ti + 2H_2O$	−860
$Zn^{2+} + 2e \rightarrow Zn$	−762.8
$ZrO_2 + 4H^+ + 4e \rightarrow Zr + 2H_2O$	−1430

REFERENCE

1. R. C. Weast, *Handbook of Chemistry and Physics*, The Chemical Rubber Company, Cleveland, Ohio, 1970.

APPENDIX D
TABLE OF SYMBOLS

	Upper Case		Lower Case		Greek
A	Amplitude, signal, area	a	Activity	α	Degree of backmixing, apex angle
A	Acid				
B	Signal	b	Bias, deadband	γ	Activity coefficient
B	Base				
C	Signal, product	c	Controlled variable		
D	Derivative time	d	Differential	Δ	Difference
E	Potential	e	Error, 2.718		
E_0	Potential at unit activity	e	Electron		
F	Flow, Faraday's constant	f	Fractional flow	π	3.1416
G	Gain	g	Vector	ρ	Density
H	Hysteresis band	h	Head	Σ	Sum
J	Integer			τ	Time constant, period
K	Equilibrium constant	k	Reaction rate coefficient, constant	τ_o	Natural period
M	Molarity, molecular weight	m	Manipulated variable	ϕ	Phase angle
N	Normality	n	Number		
P	Proportional band	p	Power		Script
		q	Process load	$\mathcal{R}$	Rangeability
R	Gas constant, reset time, resistance	r	Set point		
T	Absolute temperature	t	Time		
V	Volume				
		w	Width		
X	Reagent flow	x	Concentration		
		y	Sensitivity coefficient		
Z	Impedance, zero adjustment				

INDEX

Absorption, of ammonia, 222
  of sulfur dioxide, 88-92
Accuracy, in blending operations, 86
  in control, 183, 184
  of electrodes, 20-23
  of flowmeters, 217
Acetic acid, ionization of, 58
  titration curves, 61, 65
Activity, 1-6
  coefficient, 1, 2
  of standard solutions, 45
Adaptive control, 232-241
  by feedback, 236-241
  by feedforward, 232-236
  stability, 240, 241
Agitation, *see* Mixing
Agitator selection, 160-162
Alcohols, ionization of, 77, 78
Alkalinity of water, 70, 71
Aluminum-ion precipitation, 72
Ammonia, absorption of, 222
  anhydrous, cost of, 75
    ionization of, 77
    as a reagent, 74, 75
      in nonaqueous media, 78
  aqueous, ionization of, 59, 60
Ammonia electrode, 17, 18
Ammonium-ion electrode, 17
Ammonium nitrate, 222
Ammonium sulfide, 67
Antimony electrode, 120, 121
Argenticyanide complex ion, 99

Backmixing, 156-162
Baffles, in mixed vessels, 160

Barium carbonate, 176
Barium-ion measurement, 23, 87
Batch neutralization, 210-213
  with lime, 168
Bicarbonate alkalinity, 70, 71
Blenders, in-line, 156
Blending, of reagent and influent, 155, 156
Blending operations, 85-92
Boiler feedwater, 82
Boric acid, 78
Boron trichloride, 78
Bromide-ion electrode, 19
Buffering, 63, 64
  variable, 236, 237
Buffer solutions, 14, 44, 45
Burst cycling, 240, 241

Cadmium-ion electrode, 97
Calcium bicarbonate, in alkaline water, 71
Calcium carbonate, in alkaline water, 70, 71
  as a reagent, 176
  solubility in water, 75
Calcium fluoride, 75, 76
Calcium hydroxide, applications of, 74-77
  cost of, 75
  ionization, 74
  normality, 53
  reaction rate of, 152, 167, 168
  in scrubbing solutions, 88-92
  solubility, 74
  see also Lime
Calcium-ion electrode, 19, 20, 23
Calcium ions, in hard water, 87, 88
Calcium oxide, 175
Calcium phosphates, 76, 77

Index

Calcium salts, solubility of, 74-77
Calcium sulfate, 47, 75
Calcium sulfite, 88-92
Calibration procedures, 43-46
Calomel electrodes, 12, 13
Capacity, definition, 133
Carbonate alkalinity, 70, 71
Carbon dioxide, titration curve, 70
Carbon dioxide electrode, 17, 18
Chromate reduction, 113-119
Chromic acid, 115
Chromium ion, solubility of, 72
Chloride-ion electrode, 19
   as a reference, 122
Chloride-ion mobility, 13
Chlorinator, 112
Chlorine, in bleach production, 122, 123
   hydrolysis of, 108, 109
   as an oxidant, 107
Cleaning of electrodes, 46, 47
Closed-loop response, 201-207
Coating of electrodes, 46, 47
Common-ion effect, 96-98
Companion-ion control, 96-98
Complexation, 99, 100
Concentration versus pH, for strong acids and bases, 54, 55
   for weak acids and bases, 57-60
Controlled variable, definition, 128
Control modes, 145
   derivative, 146
   proportional, 145
   reset, 145
   windup, 214
Controllers, 145-152
   nonlinear, 194-199
      unsymmetrical, 198, 199
   proportional feedback, 191
      feedforward, 227, 228
   proportional-plus-derivative, 211, 212
   proportional-plus-reset, 149, 150
   sampled data, 199, 200
   three-mode, 146-150
Copper-ammonium ion, 99
Copper recovery, 82
Crossover period, 240
Cupric-ion electrode, 19
Curve characterizers, 193, 194
   in feedforward systems, 223, 224
Cyanate oxidation, 107, 110-113

Cyanide-ion electrode, 19
Cyanide oxidation, 107-113
Cyanogen chloride, 109, 110

Damping, in feedback loops, 129
   quarter-amplitude, 130, 131
   in sampled-data systems, 200
Dead time, 132
   as a function of flow pattern, 170
   in a mixed vessel, 159
   and mixing power, 161
   in reagent delivery, 174, 175
   sampled-data control of, 199, 200
   due to sampling, 165, 171
   two-mode control of, 149, 150
Decanting, 172
Deionized water, resistance of, 27
Delay, *see* Dead time
Derivative action, 146
Dessicants in electrode assemblies, 39, 40
Differential-ion measurement, 85
Discriminator, 238-240
Dissociation, *see* Ionization
Distilled water, resistance of, 27
Disturbance, *see* Load
Divalent-cation electrode, 23, 87, 88
Divider, as nonlinear compensator, 193

Electrical leakage, 26-33
Electrodes, *see* Inert electrodes; Reference electrode; *or* Ion-selective electrodes
Electrode assemblies, 37-43
   flow-through, 41-43
   submersible, 37-40
Electrode potential, 3
Electrode response, clean, 165
   fouled, 46
Electromotive series, 103
Emergency capacity, 172, 174
Equivalence, in reduction reactions, 104, 105
   in solubility titrations, 94
   for weak acids and bases, 62
Equivalent circuit of measuring system, 26-28
Errors in potential measurement, due to asymmetry, 36, 44
   due to fouling, 46, 47
   due to grounding, 29-33
   due to interference, 22, 23

Index

due to resistance, 26-29
due to temperature, 33-36
Extracting with water, 172

Failure modes in measuring systems, atmospheric leakage, 38-40
  electrical leakage, 26-29
  ground paths, 29-33
  solution leakage, 38-40
Failure protection, 172-174
Farraday's constant, 3
Feedback, definition, 127
Feedback adaptation, 236-241
Feedback control, 127-132, 182ff
  in feedforward systems, 228-232
Feedforward adaptation, 232-236
Feedforward control, 215-232
  adjustment of, 229, 230
  narrow-range, 221-223
  with pumped influent, 227
  with two reagents, 231
  wide-range, 223-228
Ferric ion, hydrolysis of, 73
  titration curve, 73
Ferricyanide complex ion, 100
Ferrocyanide complex ion, 100
  oxidation of, 108
Ferrous-ferric half cell, 5
Ferrous ion, in chromate reduction, 115, 117
  in hard water, 23, 87
  hydrolysis of, 72
  measurement of, 23, 87
  titration curve, 73
First-order lag, 134, 135
  step response of, 134
Flowmeters, 216-218
Flow pattern, in mixed vessels, 170
Flue-gas scrubbing, 88-92
Flumes, 218
Fluoridation of water, 86, 87
Fluoride-ion electrode, 18, 19
  interferences, 23
  temperature compensation, 35
Fluoride removal, 75, 76
Fouling of electrodes, 46, 47
Function generators, 193, 194
  in feedforward control, 224

Gain, definition, 130

of derivative control, 146
dynamic, 132-139
  of a mixed vessel, 162-164
  electrode, 140-142
  of first-order lag, 135
  of nonlinear controller, 194, 195
  of reset control, 146
  steady-state, 139-145
  of three-mode controller, 147
  of a titration curve, 140-142
  transmitter, 139
  valve, 142-145
Gas constant, 3
Glass-membrane electrodes, 15-18
  dehydration of, 29
  impedance, 28, 29
  temperature compensation, 33-35
Gold electrode, 120, 121
Grounding, 29-33

Hydrobromic acid, 57
Hydrochloric acid, activity coefficient, 2
  pH versus concentration, 54
  reagent normality, 53
Hydrofluoric acid, 66
Hydrogen chloride, 78
Hydrogen-ion electrode, 7, 102
  glass membrane, 15-17
Hydrogen-ion mobility, 13
Hydrogen sulfide, ionization of, 66, 67, 83
  stripping of, 67-69
  titration curves, 69
Hydroiodic acid, 57
Hydrolysis, of chlorine, 108, 109
  of cyanogen chloride, 109, 110
  of metal ions, 72-74
Hydroxyl-ion mobility, 13
Hypochlorite ion, reduction of, 108, 109
Hypochlorite production, 122
Hypochlorous acid, 109, 111
Hysteresis, 185

Impeller selection, 160-162
Inert electrodes, 5
  selection of, 120, 121
Interference, electrical, 31-33
  ionic, 22, 23
  in mixed-sulfide electrodes, 97
Interrupted operation, 213, 214
Iodide complex ion, 99

Index

Iodide-ion electrode, 19
Ion-exchange columns, 82
Ionic mobility, 13
Ionization, of acetic acid, 58
   of ammonia, 59, 60
   of polyprotic acids and bases, 66, 67
   of strong acids and bases, 54
   of water, 53, 54
   of weak acids and bases, 57-60
Ionization constants of acids and bases, 243-245
Ion-selective electrodes, 15-20
   glass-membrane, 15-17
      impedance, 28, 29
   liquid ion exchange, 19, 20
   mixed sulfide, 97
   permeable membrane, 17, 18
   solid-state, 18, 19
Iron metal, in copper recovery, 82
   as a reducing agent, 115
Isopotential point, 33
   of pH electrode, 34
   of fluoride-ion electrode, 35
   of silver-ion electrode, 35
   of sulfide-ion electrode, 35

K_w, 54

Lag, first-order, 134, 135
   reaction-rate, 137-139
   second-order, 138
   transportation, *see* Dead time
Laminar flow, in blending, 155
Lead-ion electrode, 97
Leak detection, with electrodes, 81, 82
Lime, applications of, 74-77
   dolomitic, 176
   feeders, 177
   handling of, 175-180
   high-calcium, 176
   reaction rate of, 152, 166-169
   slaking of, 175
   see also Calcium hydroxide
Limestone, 176. *See also* Calcium carbonate
Limit cycle, definition, 142
   in dual-reagent systems, 238, 239
   estimate of amplitude, 150
   with nonlinear controller, 196-198
   below valve throttling range, 143

Liquid junction, 6, 8
   potential, 13, 14
   resistance, 27, 33
Lithium borohydride, 78
Lithium hydroxide, 57
Lithium methylate, 78
Load, definition, 128
   response, 201-207
Loop gain, 130-132
   in sampled-data systems, 200
   and valve size, 184
Loop seals, in reagent lines, 174, 175

Magnesium hydroxide, 176
Magnesium-ion measurement, 23, 87
Magnesium oxide, 176
Magnetic flowmeter, 216, 217
Maintenance of electrodes, 43-47
Measurement location, 171
Mercuric sulfide, 98
Mercury precipitation, 97, 98
Metering pumps, in feedforward control, 223
   rangeability of, 184
   use in slurries, 180
Methanol, 77
Methyl borate (tri), 78
Mixer, in-line, 156
   selection, 160-162
   static, 156
Mixing, 155-164
   coefficient, 157, 158, 162
   with compressed air, 162
   flow patterns in, 170
   mathematical model, 157-160
   power requirements, 161
   step response of, 159, 160
Mixtures of strong and weak acids, 64, 66
   in feedforward control, 219-221, 228-232
Moisture in electrode assemblies, 30, 38-40
Molybdenum, 120
Monitoring, with ion-selective electrodes, 80-85
Multiplier, in feedforward systems, 223, 234

Nernst Equation, 2-6
   for hydrogen electrode, 102
   for reduction reactions, 104, 106
Neutrality, 53, 54

Index

Neutralization, batchwise, 210-213
  sensitivity of, 183, 184
  stagewise, 168, 169, 209, 211
  of strong acids and bases, 56
  of weak acids and bases, 60-62
Nickle-ion measurement, 23, 87
Nitrate-ion electrode, 81
Nitric acid, 57, 221, 222
Noble-metal electrodes, 120
Nonaqueous media, 77-79
Nonlinear controller, 194-199
  in adaptive systems, 234-240
  unsymmetrical, 198, 199
Nonlinear function generators, 193, 194
  in feedforward systems, 223, 224
Normal gallons-per-minute, 142
Normality, 52
  of common reagents, 53
  in feedforward control, 218-221
  versus weight percent, 52, 53

Offset, with proportional control, 145
On-off control, 213
Organic solutions, 77-79
Orifice, for blending, 155
Orifice meter, 217
ORP, 101
Oscillation, 128-132
Overload protection, 172-174
Oxidation, *see* Reduction-oxidation potential
Oxidation-reduction potential, *see* Reduction-oxidation potential
Oxidizing agents, 102-103

p notation, 4
Particle size, in reagent slurries, 176
Penalties, 182, 183
Perchloric acid, 57
Performance criteria, 200, 201
Period of oscillation, 129
  in adaptive control, 239-241
  with hysteresis, 185
  with reset, 197, 198
  variable, due to electrode fouling, 152
    due to nonlinear controller, 197, 198
    due to reaction rate, 150
Periodic disturbances, 201-207
pH, definition, 4
pH-to-current converter, *see* Transmitters

Phase shift, 128, 129
  of dead time, 133
  of derivative control, 146
  of first-order lag, 135
  of hysteresis, 185
  of reset control, 146
  of three-mode controller, 147
Phosphate removal, 76, 77
Pickle liquor, 198, 199
pIon versus activity, 4
Piping practices, 174-180
  for slurries, 177-180
Platinum electrode, 120, 121
Plug flow, 156
Polarity of electrodes, 3, 37
pH, 34, 37
Polyprotic acids and bases, 66-77
Potassium chloride, activity coefficient, 2
  solubility of, 10-12
Potassium hydroxide, 57
Potassium-ion, electrode, 17
  mobility, 13
Precipitation, of calcium salts, 74, 77
  double, 95, 96
  of metal hydroxides, 72-74
Pressurizing electrode assemblies, 42, 43
Proportional control, in feedforward systems, 227, 228
  for trim valves, 191
Proportional offset, 145
Protection against failure, 172-174
Purging electrode assemblies, 40

Quicklime, 175

Rangeability, of control valves, 143
  ball, 190
    restricted, 192
    Saunders, 180
    sequenced, 187-189
  in dual-reagent systems, 190
  of flowmeters, 216, 217
  of metering pumps, 184
  requirements, 184-186
  of solids feeders, 177
Reaction rate, 137-139
  of chromate reduction, 117
  of cyanogen chloride hydrolysis, 110
  of lime neutralization, 152
  variable, 150

258   Index

and vessel selection, 166-169
Reagents, accuracy requirements, 183, 184
  costs of, 75
  delivery, 174-180
  entry, 170
  insoluble, 175-180
  normality of, 53
  solid, 176, 177
  soluble, 174, 175
Recycling waste, 180
Redox measurement, 101ff
Reducing agents, 102, 103
Reduction-oxidation potential, 101, 102
Reduction potentials, abbrev. table, 103
  complete table, 248-250
Reduction reactions, 101
Reference electrode, 6
  calomel, 12, 13
  hydrogen, 7
  resistance of, 27
  silver, 122, 123
  silver-silver chloride, 7-12
  solid-state, 12
  special, 122, 123
  thalamid, 13
Reference solution, custom, 123
  flow rates of, 38
  potassium chloride, 9-12
  pressurization of, 42, 43
Reset action, 145
  in adaptive control, 240, 241
  in trim controllers, 191
Reset windup, 213, 214
Residence time, 159
  for chromate reduction, 117
  for cyanide oxidation, 110
  selection of, 164-169
Resistances, in measuring circuit, 26-29
Response of electrodes, clean, 165
  fouled, 46

Salts, as acids and bases, 69-74
Salt bridge, 14, 85
Sampled-data control, 199, 200
Sampling, 41-43
Saunders valve, 180
Scrubbing solutions, 88-92
Sensitivity coefficient, 204, 205
Sensitivity of electrodes, 22
Setpoint, definition, 128

Shielded wiring, 29-33
Short circuiting, in vessels, 170
Silver bromide, 93-96
Silver chloride, solubility product, 9
  solubility in KCl solution, 11
  titration, 95, 96
Silver electrode, with chlorine, 120-123
Silver-ion electrode, 9, 19
  temperature compensation, 35, 36
Silver-silver chloride electrode, 7-12
  potentials, 10
  temperature limits, 11, 12
Silver-silver chloride half cell, 3
Silver sulfide, 96, 97
Silver-sulfide electrode, 18, 19, 96, 97
Slaked lime, *see* Lime
Sludge as a buffer, 74
Slurry piping, 177-180
Slurry preparation, 180
Smoothing vessels, 170, 171
  effectiveness of, 208
Sodium acetate, 62
Sodium bisulfite, as a reducing agent, 115, 116
  in absorption of sulfur dioxide, 88-92
Sodium carbonate, 75
Sodium cyanide, 107, 108
Sodium dichromate, 115, 116
Sodium hydrosulfide, 98
Sodium hydroxide, cost, 75
  normality, 53
  pH versus concentration, 54
Sodium-ion electrode, 17
  in boiler feedwater, 82
  in scrubbing solutions, 88, 91
Sodium metabisulfite, 115, 116
Sodium sulfite, 115, 116
Solid reagent feeding, 177
Solid-state electrodes, ion-selective, 18, 19
  reference, 12
Solubility product, 75
Solubility product constants, 246, 247
Solubility titration curves, 92-96
Solution grounding, 31-33
Solution resistance, 27
Solvents, pH measurement in, 77-79
Sour water, stripping of, 67-69
Stability, 127, 128
  with adaptive controller, 240, 241
  with nonlinear controller, 196-198

Index 259

requirements, 200
Standard Hydrogen Electrode (SHE), 7
　Nernst equation for, 102
Standard Reduction Potentials, 248-250
Standard solutions, 43-46
　table of, 45
Standardization, 43-46
Step response, of a backmixed vessel, 159, 160
　of batch neutralization with lime, 168
　of a first-order lag, 134
Strong acids and bases, 51-57
Sulfide, total, determination of, 83-85
Sulfide-ion electrode, 18, 19
　differential, 85
　in mercury precipitation, 97, 98
　temperature compensation, 35, 36
Sulfur dioxide, absorption, 88-92
　in chromate reduction, 115, 116
Sulfuric acid, ionization of, 65, 66
　normality, 53
Switching of electrodes, 47
Symbols, table of, 251
Systems engineering, 182

Temperature compensation, 33-36
　for filled electrodes, 33-35
　for solid-state electrodes, 35, 36
Temperature limits, for pH electrodes, 12
　for silver-silver chloride electrode, 11-13
　for thalamid electrode, 13
Thalamid electrode, 13
Time constant, definition, 134
　of electrodes, 165
　of a mixed vessel, 160
Titration curve, for acetic acid, 61, 65
　　slope of, 64
　for alkaline water, 70
　for carbonates, 70
　for chromate reduction, 118
　for cyanide oxidation, 111
　gain of, 140-142
　of hydrogen sulfide, 69
　nonlinear properties of, 186, 187
　for reduction reactions, 105
　for silver bromide, 94
　for silver chloride, 96
　for strong acids and bases, 140, 187

　　slope of, 56, 57
　for weak acids and bases, 63, 64
Titrators, 219, 236
Tracing discharges, 81
Transmitters, 25-37
　calibration, 36, 37, 43
Transport delay, *see* Dead time
Tungsten, 120
Turbulent flow, in blending, 155

Ultrasonic cleaning of electrodes, 46

Valve, ball, 144, 179, 180
　butterfly, 144
　Saunders, 179, 180
Valve characteristics, 142-144
　ball, 179
　compensation for, 191-194
　　in feedforward systems, 233-235
　equal-percentage, 143
　　in sequenced arrangements, 187-190
　for weak acids and bases, 144
　in feedforward control, 226
　Saunders, 179
Valve characterizers, 193, 194
Valve gain, 142-145
　adaptation for, 233-235
　equal-percentage, 143
Valve positioners, 189
　nonlinear, 193
Valve rangeability, 143
　in feedforward control, 224-228
Valve selection, 164-174
　for slurries, 179, 180
Valve sequencing, 187-190
Vessels, smoothing, 170, 171
Vessel selection, 164-174
Vortex elimination, 162

Water, ionization of, 53, 54
　resistance of, 27
Water-hardness electrode, 23, 87, 88
Weak acids and bases, 57-74
Weirs, 217, 218
Wiring practices, 29-33, 37, 38

Zinc-ion measurement, 23, 87
Zinc-ion precipitation, 71, 72